Perspectives in Phytochemistry

Perspectives in Phytochemistry

PROCEEDINGS OF THE
PHYTOCHEMICAL SOCIETY SYMPOSIUM
CAMBRIDGE, APRIL 1968

Edited by

J. B. HARBORNE

Department of Botany
University of Reading, Reading, England

AND

T. SWAIN

Department of Biology
Yale University, New Haven, Connecticut, U.S.A.

1969

ACADEMIC PRESS
LONDON AND NEW YORK

ACADEMIC PRESS INC. (LONDON) LTD.
Berkeley Square House
Berkeley Square
London, W1X 6BA

U.S. Edition published by
ACADEMIC PRESS INC.
111 Fifth Avenue
New York, New York 10003

Library of Congress Catalog Card Number: 69-16496

PRINTED IN GREAT BRITAIN BY
SPOTTISWOODE, BALLANTYNE & CO. LTD.
LONDON AND COLCHESTER

Contributors

E. C. BATE-SMITH, *Agricultural Research Council Institute of Animal Physiology, Babraham, Cambridge, England.*

G. W. BUTLER, *Plant Chemistry Division, Division of Scientific and Industrial Research, Palmerston North, New Zealand.*

E. E. CONN, *Department of Biochemistry and Biophysics, University of California at Davis, Davis, California, U.S.A.*

H. ERDTMAN, *Department of Organic Chemistry, Royal Institute of Technology, Stockholm, Sweden.*

A. W. GALSTON, *Department of Biology, Yale University, New Haven, Connecticut, U.S.A.*

T. W. GOODWIN, *Biochemistry Department, University of Liverpool, Liverpool, England.*

R. HEGNAUER, *Department of Experimental Plant Systematics, University of Leiden, Netherlands.*

V. HEROUT, *Czechoslovak Academy of Sciences, Prague, Czechoslovakia.*

A. T. JAMES, *Unilever Research Laboratories, Sharnbrook, Bedford, England.*

T. J. MABRY, *The Cell Research Institute and Department of Botany, The University of Texas at Austin, Austin, Texas, U.S.A.*

F. ŠORM, *Czechoslovak Academy of Sciences, Prague, Czechoslovakia.*

G. H. N. TOWERS, *Department of Botany, University of British Columbia, Vancouver, Canada.*

Preface

To celebrate its Tenth Anniversary, the Phytochemical Society held a three-day Symposium at Cambridge in April 1968 at which leading authorities reviewed the development of phytochemistry over the last ten years. It was clearly impossible, in the time available, to touch upon the highlights in more than a few fields, such as the structural elucidation, biosynthesis, chemotaxonomy, metabolism and function of selected groups of plant constituents. A representative coverage of these different topics is presented in this book.

Following the first chapter which illustrates the important contribution of physical methods in the structural elucidation of natural products, there are three chapters devoted to the biosynthesis of various types of secondary plant constituents. The following four chapters deal with chemotaxonomy, a subject which has developed into a major topic during the period under review. It is particularly pleasing to include a lecture, reproduced here as Chapter 8, given at the Symposium by the founder of the Society, Dr. E. C. Bate-Smith, C.B.E. The final two chapters are devoted to the metabolism and function of two related groups of plant substances.

Although four chapters deal with phenolic compounds, such a bias is not inappropriate in view of the origin of the Society as the Plant Phenolics Group. Other classes of natural constituents which are discussed, include carotenoids, steroids, fatty acids, polyacetylenes, iridoids, cyanogenic glycosides and sesquiterpene lactones. In presenting general reviews of their chosen subjects the authors have taken great care to be up to date and, in many cases, have included recent unpublished results from their own laboratories. We believe, therefore, that the book will be of great interest to all plant scientists.

In conclusion we would like to express our thanks to Lord Todd, who gave the Opening Address at the Symposium, and to all our other distinguished colleagues who chaired the different sessions and helped in other ways to make the Symposium a success. We are also very grateful to the contributors, who have responded generously and promptly to the various demands made on them, and to the staff of Academic Press for their expert assistance in preparing the book for publication.

December, 1968

J. B. HARBORNE
T. SWAIN

Contents

List of Contributors v
Preface vii

CHAPTER 1

The Ultraviolet and Nuclear Magnetic Resonance Analysis of Flavonoids

T. J. Mabry

I. Introduction 1
II. The Structure Analysis of Flavonoids by Ultraviolet Spectroscopy . 2
III. Flavonoid Structure Analysis by Nuclear Magnetic Resonance Spectroscopy 31
Appendix 43
References 44

CHAPTER 2

The Biosynthesis of Cyanogenic Glycosides and Other Simple Nitrogen Compounds

Eric E. Conn and G. W. Butler

I. Introduction 47
II. Early Studies on Biosynthesis 50
III. Recent Studies on Biosynthesis 57
IV. Biosynthesis of Mustard Oil Glucosides . . . 63
V. Metabolism of Cyanide 66
VI. Conclusion 70
References 72

CHAPTER 3

Recent Investigations on the Biosynthesis of Carotenoids and Triterpenes

T. W. Goodwin

I. Introduction 75
II. Stereospecific Biosynthesis of Squalene and Phytoene . . 75
III. Mechanism of Formation of Cyclic Carotenes . . 78
IV. Formation of Phytoene and its Subsequent Desaturation . 82
V. Phytoene Synthesis by Isolated Chloroplasts . . . 84
VI. The Mechanism of Formation of Cycloartenol . . 85
Acknowledgements . . , ... 89
References 89

CHAPTER 4

Fatty Acid Biosynthesis in Plants

A. T. James

I.	Saturated Fatty Acid Biosynthesis	91
II.	Unsaturated Fatty Acid Biosynthesis	93
III.	Unusual Fatty Acids	102
IV.	The Effect of Temperature on Fatty Acid Biosynthesis	104
References		105

CHAPTER 5

Recent Developments in Molecular Taxonomy

H. Erdtman

I.	Introduction	107
II.	Fossils Ancient and "Modern"	108
III.	Comparative Phyto- and Zoochemistry	109
IV.	Biosynthesis and Molecular Taxonomy	112
V.	Molecular Biology and Molecular Taxonomy	118
VI.	The Future	118
References		119

CHAPTER 6

Chemical Evidence for the Classification of some Plant Taxa

R. Hegnauer

I.	Introduction	121
II.	Phytochemistry and Plant Classification	123
III.	Comparative Studies and Classification of Taxa	134
IV.	Conclusion	135
Acknowledgements		136
References		136

CHAPTER 7

Chemotaxonomy of the Sesquiterpenoids of the Compositae

V. Herout and F. Šorm

I.	Introduction	139
II.	The Definition of Sesquiterpene Lactones as Taxonomic Characters	140
III.	The Probable Biosynthetic Pathway to Sesquiterpene Lactones	142
IV.	Sesquiterpene Lactones and the Classification of the Compositae into Tribes	144
V.	Sesquiterpene Lactones and the Classification within Tribes	154
VI.	Limitations in the Taxonomic Significance of the Phytochemical Data	161
VII.	Conclusions	162
References		163

CHAPTER 8

Flavonoid Patterns in the Monocotyledons

E. C. Bate-Smith

I.	Introduction	167
II.	Flavonoid Patterns in the Dicotyledons	169
III.	Monocotyledons and Dicotyledons Compared	172
IV.	Flavonoids of Selected Monocotyledonous Families	172
V.	Taxonomic and Phylogenetic Implications of the Flavonoid Patterns	175
References		177

CHAPTER 9

Metabolism of Cinnamic Acid and its Derivatives in Basidiomycetes

G. H. N. Towers

I.	Introduction	179
II.	Phenylalanine and Tyrosine Ammonia Lyases in Basidiomycetes	181
III.	Metabolism of Phenylalanine in Basidiomycetes	182
IV.	Degradation of Phenylalanine and Tyrosine in Basidiomycetes	183
V.	Formation of more complex Cinnamoyl Derivatives including Lignins	187
VI.	Comparison of Cinnamate Metabolism in Basidiomycetes and in Higher Plants	189
References		190

CHAPTER 10

Flavonoids and Photomorphogenesis in Peas

A. W. Galston

I.	Introduction	193
II.	Growth Responses of the Pea to Light	194
III.	Identification of Flavonoids Affecting IAA Oxidase Activity	196
IV.	Relation of Flavonoid Content to Growth	197
V.	Effect of Light on Uptake of Flavonoid Precursors	198
VI.	Flavonoids in Pea Tendrils	200
VII.	Other Relevant Recent Investigations	203
VIII.	Conclusion	203
References		204

Author Index	205
Chemical Compounds Index	213
Genus and Species Index	221
Subject Index	227

CHAPTER 1

The Ultraviolet and Nuclear Magnetic Resonance Analysis of Flavonoids

TOM J. MABRY

*The Cell Research Institute and Department of Botany,
The University of Texas at Austin, Austin, Texas, U.S.A.*

I. Introduction	1
II. The Structure Analysis of Flavonoids by Ultraviolet Spectroscopy . .	2
A. The UV Spectra of Flavones and Flavonols	2
B. The UV Spectra of Isoflavones, Flavanones and Dihydroflavonols in Methanol	20
III. Flavonoid Structure Analysis by Nuclear Magnetic Resonance Spectroscopy	31
A. Introduction	31
B. The Use of DMSO-d_6 as a Solvent for Flavonoid NMR Spectroscopy	31
C. Preparation of Trimethylsilyl Ether Derivatives of Flavonoids . .	32
D. Interpretation of the NMR Spectra of Fully Trimethylsilylated Flavonoids	32
Appendix	43
References	44

I. INTRODUCTION

During the last ten to fifteen years, ultraviolet (UV) and nuclear magnetic resonance (NMR) spectroscopy have been developed by chemists into routine techniques for the identification and structure determination of flavonoids. Nevertheless, many biologists still experience considerable difficulty in using these spectral methods of flavonoid analyses; in part, because much of the available data is either incomplete or was not determined under standard conditions. In an attempt to alleviate this problem, we (Drs K. R. Markham, M. B. Thomas and I) began in 1965 to collect all the data which we had found useful in our laboratory for flavonoid identification into a single compilation entitled "The Systematic Identification of Flavonoids" (Mabry *et al.*, 1969). This book is divided into three parts; the first deals mainly with the isolation and purification of flavonoids, while the other two sections deal with the UV and NMR spectra of these compounds. The information presented in these latter two sections are summarized in this chapter.

Although we concentrated on the spectral analysis of flavones, isoflavones and flavonols, our investigation included data for a few selected flavanones, dihydroflavonols, chalcones and aurones. The latter two classes of flavonoids, however, are not covered here.

II. THE STRUCTURE ANALYSIS OF FLAVONOIDS BY ULTRAVIOLET SPECTROSCOPY

Our data complement previous investigations and reviews of the UV spectral analysis of flavonoids such as that of Jurd (1962), who reviewed the literature prior to 1962 and assembled the UV spectral data available at that time. Each of the 175 flavonoids examined in our study is represented by a set of six ultraviolet spectra, all of which are presented in the compilation by Mabry et al. (1969) (Fig. 1). Each set of spectra consists of one determined on a solution of the flavonoid in methanol and five obtained by adding diagnostic reagents to the methanol solution of the flavonoid. The reagents are sodium methoxide, aluminium chloride, aluminium chloride–hydrochloric acid, sodium acetate and sodium acetate–boric acid. Details of these procedures are given in the Appendix.

A. THE UV SPECTRA OF FLAVONES AND FLAVONOLS

UV spectroscopy is particularly applicable to flavones (I) and flavonols (3-hydroxyflavones; II) because of the direct conjugation of both the A and B rings to the carbonyl group.

(I) Flavone skeleton   (II) Flavonol skeleton

The UV spectral data for a selected group of flavones and flavonols are presented in Table I.

1. The UV Spectra of Flavones and Flavonols in Methanol

The methanol spectra of flavones and flavonols exhibit two major absorption peaks in the region 240–400 nm (mμ). These two peaks are commonly referred to as Band I (usually 300–380 nm) and Band II (usually 240–280 nm) (Fig. 2). Band I is considered to be associated with absorption due to the B-ring

7-HYDROXYFLAVONE

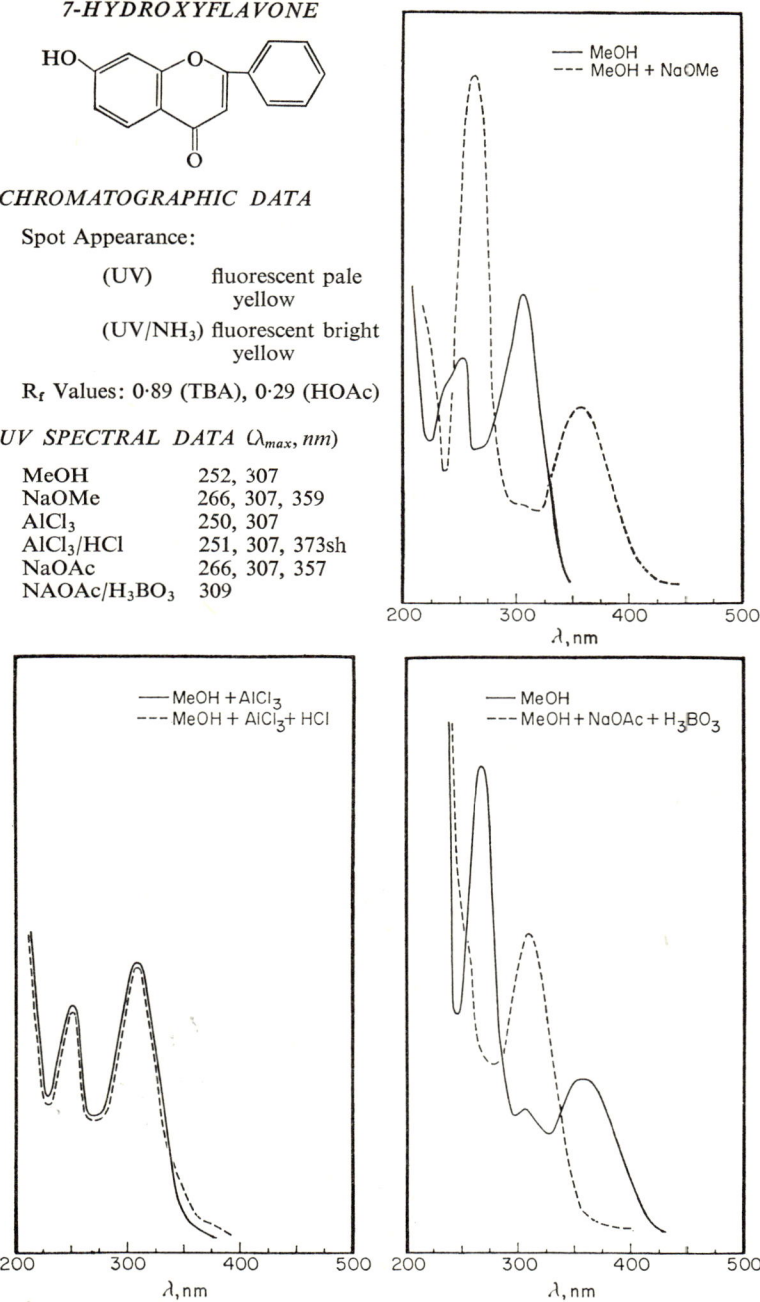

CHROMATOGRAPHIC DATA

Spot Appearance:

(UV)	fluorescent pale yellow
(UV/NH$_3$)	fluorescent bright yellow

R$_f$ Values: 0·89 (TBA), 0·29 (HOAc)

UV SPECTRAL DATA (λ_{max}, nm)

MeOH	252, 307
NaOMe	266, 307, 359
AlCl$_3$	250, 307
AlCl$_3$/HCl	251, 307, 373sh
NaOAc	266, 307, 357
NAOAc/H$_3$BO$_3$	309

Fig. 1. The UV spectra of 175 flavonoids have been recorded elsewhere (Mabry *et al.* 1969) in the form shown here for one of them, 7-hydroxyflavone.

TABLE I

The ultraviolet spectra[a] of flavones and flavonols

Flavonoid	MeOH λ_{max} (nm)	R.A.[a]	NaOMe λ_{max} (nm)	R.A.	AlCl$_3$ λ_{max} (nm)	R.A.	AlCl$_3$/HCl λ_{max} (nm)	R.A.	NaOAc λ_{max} (nm)	R.A.	NaOAc/H$_3$BO$_3$ λ_{max} (nm)	R.A.
1. Flavone	294	1·0	294	1·0	294	1·0	294	1·0	293	1·0	295	1·0
	250	0·8	250	0·8	250	0·8	250	0·8	248	0·9	256sh	0·8
2. 3′,4′-Dihydroxyflavone	342	1·0	404	1·0	468sh	1·0	342	1·0	400	1·0	365	1·0
	311sh	0·7	301	0·5	379	3·3	312sh	0·7	347	0·9	306	0·8
	243	0·8	276sh	0·4	306	2·6	244	0·8	306sh	0·9		
					273sh	1·8						
					248sh	3·4						
3. 4′,7-Dihydroxyflavone	328	1·0	386	1·0	383sh	1·0	395	1·0	368	1·0	329	1·0
	253sh	0·4	329	0·4	328	3·2	328	2·0	320sh	0·8	314sh	0·9
			244sh	0·5	313sh	2·9	310sh	1·8	309	0·8	256sh	0·5
			252	0·6	255sh	1·8	248sh	1·5	261	0·9		
4. Bayin (8-C-glycosyl-4′,7-dihydroxyflavone)	328	1·0	390	1·0	284sh	1·0	398	1·0	369	1·0	331	1·0
	255sh	0·5	333	0·4	331	3·4	330	1·5	309	0·9	315sh	0·9
			267	0·6	313	3·2	310	1·5	268	1·3	257sh	0·6
			255	0·6	254sh	2·3	252sh	1·3				
5. 3′,4′,7-Trihydroxyflavone	343	1·0	395	1·0	458	1·0	409	1·0	373	1·0	361	1·0
	309	0·9	338sh	0·5	371	1·3	339	1·8	310sh	0·8	306	0·7
			314sh	0·4	304	1·3	307	1·6	254	1·6		
			256	1·2								

6. Baicalein (5,6,7-trihydroxyflavone)	323	1·0	(dec.) 420sh	1·0	375	1·0	346	1·0	(dec.) 362	1·0	334	1·0
	274	1·9	366	2·6	314sh	0·5	293sh	1·2	258	2·4	277	1·6
	247sh	1·0	257	2·0	282sh	1·2	282	1·4				
					272	1·3	255sh	0·7				
					247	1·0						
7. Norwogonin (5,7,8-trihydroxyflavone)	350sh	1·0	(dec.)		362sh	1·0	343sh	1·0	(dec.)		287	1·0
	280	4·4			314	4·3	296	2·7	274	1·0		
	235	5·7			289sh	4·0	236	4·1				
					238	8·2						
8. Apigenin (4′,5,7-trihydroxyflavone)	338	1·0	392	1·0	384	1·0	381	1·0	376	1·0	344	1·0
	297sh	0·7	324	0·5	348	1·1	341	1·4	301	0·8	302sh	0·7
	267	0·9	275	0·8	301	0·9	300	1·1	274	1·3	268	1·0
					276	1·0	276	1·1				
9. Apigenin 7-O-glucoside	333	1·0	388	1·0	383	1·0	381	1·0	387	1·0	340	1·0
	268	1·0	295sh	0·2	346	1·3	341	1·4	355	1·0	267	0·9
			267	0·5	299	0·8	299	1·1	267	1·0		
					276	1·1	277	1·3	256sh	0·9		
10. Isovitexin (6-C-glucosylapigenin)	336	1·0	398	1·0	382	1·0	380	1·0	384	1·0	345	1·0
	271	0·9	330	0·4	352	1·2	344	1·4	303	0·8	272	1·1
			278	0·6	304	0·8	301	1·1	278	1·3		
					278	1·0	280	1·2				
11. Acacetin (apigenin 4′-methyl ether)	327	1·0	364	1·0	382	1·0	379	1·0	357	1·0	331	1·0
	268	1·0	295sh	1·4	344	1·3	338	1·6	294sh	1·5	269	1·1
			276	2·2	302	1·2	300	1·5	276	2·1		
					277	1·2	279	1·5				

TABLE I (*continued*)

6

Flavonoid	MeOH λ_{max} (nm)	R.A.[a]	NaOMe λ_{max} (nm)	R.A.	AlCl$_3$ λ_{max} (nm)	R.A.	AlCl$_3$/HCl λ_{max} (nm)	R.A.	NaOAc λ_{max} (nm)	R.A.	NaOAc/ H$_3$BO$_3$ λ_{max} (nm)	R.A.
12. Luteolin (3′,4′,5,7-tetrahydroxyflavone)	349	1·0	401	1·0	426	1·0	384	1·0	384	1·0	433sh	1·0
	291sh	0·5	330sh	0·3	329	0·2	355	1·0	327sh	0·6	371	2·8
	267	0·8	265	0·8	299sh	0·3	294sh	0·7	269	1·2	259	3·2
	253	0·9			273	0·7	275	1·0				
							266sh	1·0				
13. Isoorientin (8-*C*-glucosyl-3′,4′,5,7-tetra-hydroxyflavone)	349	1·0	407	1·0	429	1·0	385	1·0	392	1·0	430sh	1·0
	270	0·9	337sh	0·3	333	0·2	361	1·0	322	0·6	377	1·7
	255	0·8	267	0·7	302sh	0·3	296sh	0·7	276	1·1	265	2·0
					278	0·8	279	1·0				
							265sh	0·9				
14. Chrysoeriol (3′-methoxy-4′,5,7-trihydroxy-flavone)	348	1·0	405	1·0	390	1·0	386	1·0	395	1·0	351	1·0
	267	0·8	329sh	0·3	366sh	0·9	353	1·1	320	0·6	268	0·9
	242	0·8	276sh	0·6	297	0·5	294	0·7	271	1·0		
			265	0·6	274	0·8	276	1·0				
					262	0·8	259	0·9				
15. Tricin (3′,5′-dimethoxy-4′,5,7-trihydroxy-flavone)	348	1·0	416	1·0	387sh	1·0	387sh	1·0	414	1·0	348	1·0
	268	1·0	258	0·9	359	1·0	355	1·1	264	1·6	269sh	1·4
					300sh	0·9	303sh	1·0				
					276	1·5	276	1·6				
					254sh	1·7	253sh	1·8				

16. Nevadensin (4′,6,8-trimethoxy-5,7-dihydroxyflavone)	330	1·0	377	1·0	413sh	1·0	403sh	1·0	376	1·0	409sh	1·0
	284	1·2	299sh	2·0	356	3·0	351	3·3	302sh	1·8	322sh	2·2
			283	2·5	310	3·0	309	3·6	283	2·4	286	3·1
					290sh	2·2	262	1·1				
					265sh	1·3						

17. Hymenoxin (3′,4′,6,8-tetramethoxy-5,7-dihydroxyflavone)	337	1·0	369	1·0	364	1·0	306sh	1·0	378	1·0	329	1·0
	279	1·1	311sh	1·6	290	0·9	357	4·4	312sh	1·2	281	1·3
			286	1·8			293	4·2	283	1·7		

18. Amentoflavone (a 3″, 8-linked apigenin biflavonyl)	335	1·0	382	1·0	386	1·0	385	1·0	369	1·0	332	1·0
	292sh	1·0	295sh	0·7	350	1·2	344	1·4	293sh	1·2	271	1·4
	269	1·2	275	1·1	299	1·0	299	1·2	274	1·5		
					277	1·2	278	1·4				

			(dec.)								
19. 3,4′,7-trihydroxyflavone	357	1·0	408	1·0	418	1·0	416	1·0	378	1·0	425sh	1·0
	318	0·7	327	0·5	322	0·2	324	0·2	328	0·8	357	5·0
	258	0·7	319sh	0·4	271	0·6	271	0·6	318sh	0·7	318	3·4
			275	0·6	256sh	0·5	255sh	0·5	268	1·0	259	3·8

			(dec.)								
20. Kaempferol (3,4′,5,7-tetra-hydroxyflavone)	367	1·0	416	1·0	424	1·0	424	1·0	388	1·0	372	1·0
	322sh	0·5	316	0·5	350	0·3	349	0·4	303	0·6	322sh	0·5
	266	0·8	278	0·9	304	0·2	304sh	0·3	274	1·2	297sh	0·5
					268	0·9	269	0·8			267	1·0
					260sh	0·7	258sh	0·7				

TABLE I (*continued*)

8

Flavonoid	MeOH λ_{max} (nm)R.A.[a]		NaOMe λ_{max} (nm) R.A.		AlCl$_3$ λ_{max} (nm) R.A.		AlCl$_3$/HCl λ_{max} (nm) R.A.		NaOAc λ_{max} (nm) R.A.		NaOAc/ H$_3$BO$_3$ λ_{max} (nm) R.A.	
21. Kaempferol 7-*O*-neohesperidoside	365	1·0	(dec.) 425	1·0	426	1·0	422	1·0	408sh	1·0	370	1·0
	323sh	0·5	267	0·9	352	0·4	350	0·5	385	1·2	325sh	0·5
	265	0·8	244	1·0	301sh	0·2	300sh	0·3	323	0·6	264sh	1·0
	252	0·8			267	0·9	266	1·0	260	1·7		
					259sh	0·9	258sh	0·9				
22. Robinin (kaempferol 3-*O*-robinoside 7-*O*-rhamnosylglucoside)	350	1·0	389	1·0	400	1·0	398	1·0	406sh	1·0	352	1·0
	265	1·2	269	0·9	354	0·9	348	1·1	358	1·6	265	1·2
			245	0·8	301	0·6	298sh	0·7	265	2·1		
					274	1·3	274	1·4				
23. Kaempferol 4′-methyl ether	366	1·0	411	1·0	422	1·0	421	1·0	384	1·0	367	1·0
	320	0·6	323sh	0·5	350	0·4	347	0·5	301sh	0·8	318	0·6
	267	0·9	280	1·2	305	0·3	305sh	0·3	274	1·4	268	1·1
	254sh	0·8			270	0·9	267	1·0				
24. Fisetin (3,3′,4′, 7-tetrahydroxyflavone)	362	1·0	(dec.) 341	1·0	458	1·0	422	1·0	(dec. 378	1·0	382	1·0
	319sh	0·5	292	1·3	316sh	0·2	323	0·2	330	0·6	316	0·5
	248	0·7	252	2·0	286	0·5	262	0·6	320	0·6	265sh	0·8
									263sh	0·7		

						(hydrolyzes)							
25. Fisetin 3-O-glucoside	340	1·0	408	1·0	379	1·0	420	1·0	369sh	1·0	365	1·0
	310	1·1	324	1·0	316sh	1·4	307	1·1	317	1·1	311	1·0
	255sh	1·4	256	2·0	276	1·8	272sh	1·4	256sh	1·9		
							254	1·5				

			(dec.)				(dec.)					
26. Quercetin (3,3′,4′,5,7-pentahydroxy-flavone)	369	1·0	322	1·0	458	1·0	428	1·0	390	1·0	387	1·0
	301sh	0·3			335	0·1	360	0·4	327	0·7	261	1·1
	255	1·0			273	0·8	302sh	0·2	274	1·0		
							265	1·0	256	0·9		

27. Quercitrin (quercetin 3-O-rhamnoside)	350	1·0	394	1·0	431	1·0	401	1·0	373	1·0	367	1·0
	256	1·4	327	0·5	333	0·3	353	0·8	322sh	0·7	260	1·5
			271	1·4	305sh	0·4	302sh	0·6	272	1·7		
					276	1·2	272	1·7				

28. Quercetin 3,7-O-diglucoside	356	1·0	396	1·0	440	1·0	402	1·0	422sh	1·0	380	1·0
	256	1·4	268	1·2	335	0·2	363sh	0·8	371	1·3	261	1·4
					275	1·2	300sh	0·6	260	2·3		
							270	1·6				

			(dec.)				(dec.)					
29. Rhamnetin (quercetin 7-methyl ether)	370	1·0	432	1·0	451	1·0	424	1·0	383	1·0	390	1·0
	256	1·2	332	0·6	272	0·9	364sh	0·5	255	1·5	261	1·2
			285	0·8			300sh	0·4				
			242	1·2			266	1·1				

TABLE I (continued)

Flavonoid	MeOH λ_{max} (nm) R.A.[a]		NaOMe λ_{max} (nm) R.A.		AlCl$_3$ λ_{max} (nm) R.A.		AlCl$_3$/HCl λ_{max} (nm) R.A.		NaOAc λ_{max} (nm) R.A.		NaOAc/ H$_3$BO$_3$ λ_{max} (nm) R.A.	
30. Isorhamnetin (quercetin 3′-methyl ether)	371	1·0	(dec.) 448	1·0	431	1·0	428	1·0	(dec.) 393	1·0	377	1·0
	305sh	0·3	329	2·5	361sh	0·3	357	0·4	320	0·6	254	1·1
	255	1·0	241sh	2·3	304sh	0·2	302sh	0·2	275	1·0		
					264	0·9	262	0·9				
31. Isorhamnetin 3-O-rutinoside	356	1·0	414	1·0	401	1·0	400	1·0	396	1·0	360	1·0
	306sh	0·5	330	0·3	368sh	0·9	360	0·9	320sh	0·7	305sh	0·6
	265sh	1·0	271	0·9	300sh	0·6	302sh	0·6	271	1·4	268sh	1·1
	254	1·2			267	1·4	267	1·4			254	1·2
32. Tamarixetin 7-O-rutinoside (quercetin 4′-methyl ether 7-O-rutinoside)	368	1·0	417	1·0	427	1·0	423	1·0	388	1·0	371	1·0
	255	1·2	268	1·1	365sh	0·4	359	0·5	327	0·6	255	1·3
					303sh	0·3	301sh	0·4	256	1·6		
					266	1·1	266	1·1				
33. Robinetin (3,3′,4′,5′, 7-pentahydroxy-flavone)	367	1·0	(dec.) 475	1·0	447	1·0	428	1·0	(dec.) 345	1·0	462sh	1·0
	320	0·5	333	1·6	312	0·2	317	0·2	308sh	0·5	385	5·5
	252	0·6			278	0·4	271	0·5	255	0·9	316	2·8
											256sh	4·3

No.														
34. Penduletin (3′,6,7-trimethoxy-4′,5-dihydroxyflavone)	340	1·0	390	1·0	369	1·0	359	1·0	396	1·0	343	1·0
	271	0·9	274	0·8	302sh	0·6	301sh	0·7	349	1·0	271	0·8
					280	0·7	283	0·8	273	1·7		
35. Jacein (3,3′,6-trimethoxy-4′,5,7-trihydroxyflavone 7-O-glucoside)	352	1·0	401	1·0	387	1·0	407sh	1·0	420	1·0	357	1·0
	272sh	0·8	270	0·9	270	1·0	369	1·3	380sh	0·8	271sh	0·9
	257	0·9	248	0·8			280sh	1·3	265	1·2	257	1·0
							267	1·3				
36. Centaurein (3,4′,6-trimethoxy-3′,5,7-trihydroxyflavone 7-O-glucoside)	350	1·0	381	1·0	381	1·0	402sh	1·0	348	1·0	352	1·0
	270sh	0·9	274	2·3	271	1·0	367	1·4	271	1·1	270sh	0·9
	256	1·0					281	1·4	255	1·1	258	1·1
							268	1·4				
37. Patulitrin (3,3′,4′,5,7-pentahydroxy-6-methoxyflavone 7-O-glucoside)			(dec.)						(dec.)			
	372	1·0	444sh	1·0	461	1·0	425	1·0	417sh	1·0	393	1·0
	259	1·1	386	1·3	346sh	0·2	380sh	0·7	397	1·1	263	1·2
			292	3·1	277	0·8	269	1·1	342	0·7		
			242	4·1					258	1·6		

[a] The relative absorptivities (R.A.) are presented for the λ_{max}'s obtained for each spectrum using the longest wavelength peak or shoulder as 1·0. dec. = decomposes (usually within 5 or 10 minutes).

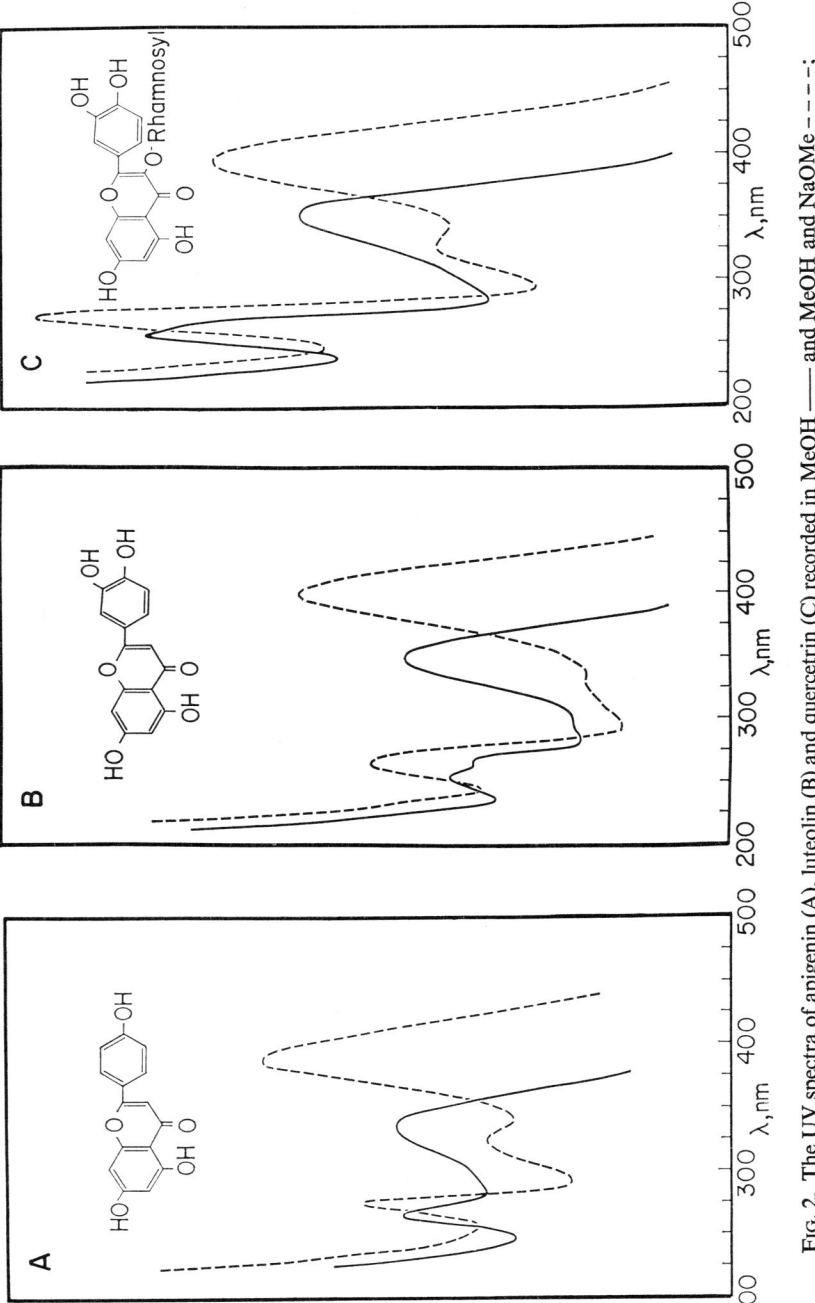

Fig. 2. The UV spectra of apigenin (A), luteolin (B) and quercetrin (C) recorded in MeOH —— and MeOH and NaOMe - - - -;

cinnamoyl system, and Band II with absorption involving the A-ring benzoyl system (see III) (Jurd, 1962).

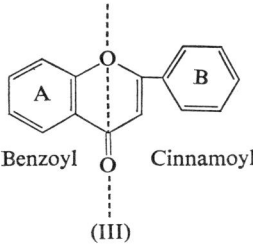

Benzoyl   Cinnamoyl

(III)

These assignments are supported, to some extent, by the UV spectral data of substituted flavonoids. For example, flavones and flavonols oxygenated in the A-ring, but not in the B-ring, tend to give spectra in methanol with a

TABLE II

Band I in the UV spectra of flavones and flavonols

Flavonoid type	Number of compounds examined	Range of Band I (nm)
Flavones	50	294–352
Flavonols (3-hydroxyl substituted)	26	330–359
Flavonols (free 3-hydroxyl)	27	354–387

pronounced Band II and a weak Band I; but in similar compounds which also possess B-ring oxygenation, Band I is more pronounced and appears at longer wavelengths. It is apparent from the data presented in Table II that the position of Band I distinguishes between flavones and flavonols (with free 3-hydroxyl groups).

2. The Effect of Oxidation Patterns on the UV Spectra of Flavones and Flavonols

On increasing the oxygenation of the B-ring in flavones and flavonols, about a 3–8 nm bathochromic shift in Band I occurs with each additional oxygen function (see Table III). On the other hand, while changes in the B-ring oxygenation pattern usually do not produce a shift in Band II, Band II may appear as either one or two peaks (designated IIa and IIb with IIa being the peak at longer wavelength) depending on the B-ring oxidation pattern. For example, the 3',4'- (and to a lesser extent the 3',4',5'-) oxygenated flavones and flavonols usually exhibit two absorption peaks between 250–275 nm, while the 4'-oxygenated equivalents have only one (compare the MeOH spectra for apigenin and luteolin, Fig. 2).

TABLE III

Band I in the UV spectra of flavonols differing in their B-ring
oxidation pattern

Flavonol	Oxidation pattern		Band I (nm)
	A- and C-rings	*B-ring*	
Galangin	3,5,7	—	359
Kaempferol	3,5,7	4'	367
Quercetin	3,5,7	3',4'	371
Myricetin	3,5,7	3',4',5'	374

Increasing hydroxylation of the A-ring in flavones and flavonols produces a notable bathochromic shift in Band II (see Table IV) and a smaller effect on Band I.

TABLE IV

Band II in the UV spectra of flavones having oxidation only in the A-ring

Flavone	A-ring pattern oxidation	Band II (nm)
Flavone	—	250
5-Hydroxyflavone	5	270
7-Hydroxyflavone	7	252
5,7-Dihydroxyflavone	5,7	268
Baicalein	5,6,7	274
Norwogonin	5,7,8	280

3. The Effect of Methylation and Glycosylation on the UV Spectra of Flavones and Flavonols

If either a 3-, 5- or 4'-hydroxyl group on the flavone or flavonol nucleus is methylated or glycosylated, hypsochromic shifts (i.e. to shorter wavelengths), especially in Band I, are observed. The shift associated with the substitution of the 3-hydroxyl is usually of the order of 12–17 nm. Methylation of the 5-hydroxyl group results in a 5–15 nm hypsochromic shift in both Band I and II, and methylation or glycosylation of the 4'-hydroxyl group produces a 3–10 nm hypsochromic shift in Band I. Substitution of any hydroxyl group other than those at positions 3, 5 and 4' has little or no effect on the UV spectrum.

4. The Effect of Acetylation on the UV Spectra of Flavones and Flavonols

Acetylation of a phenolic hydroxyl group nullifies the effect of that group on the UV spectrum. For example, diosmetin triacetate (IV) possesses a UV spectrum (λ_{max} 259 and 322 nm) similar to that observed for 4'-methoxyflavone (λ_{max} 253 and 317 nm).

(IV) Diosmetin triacetate

5. The UV Spectra of Flavones and Flavonols in the Presence of NaOMe

Sodium methoxide is a strong base and ionizes to some extent all hydroxyl groups on the flavonoid nucleus. The addition of NaOMe to flavones and flavonols in methanol usually produces bathochromic shifts in all absorption bands (Fig. 2). However, a large bathochromic shift of Band I of about 45–65 nm, without a decrease in intensity, is diagnostic for the presence of a free 4'-hydroxyl group. Flavonols lacking a free 4'-hydroxyl but having a free 3-hydroxyl group also give a 50–60 nm bathochromic shift in Band I; however there is usually a decrease in intensity of the peak.

Flavonols which have free hydroxyl groups at both the 3- and 4'-positions are unstable in NaOMe and the absorption peaks in the NaOMe spectrum degenerate in a few minutes. Flavonols which contain a 3,3',4'-trihydroxyl system decompose even faster than those having the 3,4'-dihydroxylation pattern. Although alkali instability is generally associated with flavonols having the 3,4'-dihydroxyl grouping, other hydroxylation patterns in flavones, notably 5,6,7; 5,7,8 and 3',4',5', are also alkali sensitive.

6. The UV Spectra of Flavones and Flavonols in the Presence of NaOAc

Sodium acetate is a weaker base than NaOMe, and, as such, ionizes only the more acidic hydroxyl groups in flavones and flavonols, i.e. the 3-, 7- and 4'-hydroxyl groups. Because ionization of the 7-hydroxyl group mainly affects Band II (whereas ionization of the 3- and/or 4'-hydroxyl groups mainly affects Band I), NaOAc is a particularly useful diagnostic reagent for the specific detection of 7-hydroxyl groups.

The UV spectra of flavones and flavonols containing free 7-hydroxyl groups with few exceptions exhibit a diagnostic 5–20 nm bathochromic shift of Band II in the presence of NaOAc (see Fig. 3). However, when 6,8-oxygen substituents are present in flavones, the bathochromic shift with NaOAc is often small or imperceptible.

If the NaOAc spectrum of a flavone or flavonol changes after several minutes, then the flavonoid has decomposed due to the presence of an alkali-sensitive grouping. The most common alkali-sensitive oxygenation patterns in the flavones and flavonols are those which contain 5,6,7; 5,7,8 or 3,3',4' hydroxyl groups (in the latter pattern the 3'-function may be a methoxyl group).

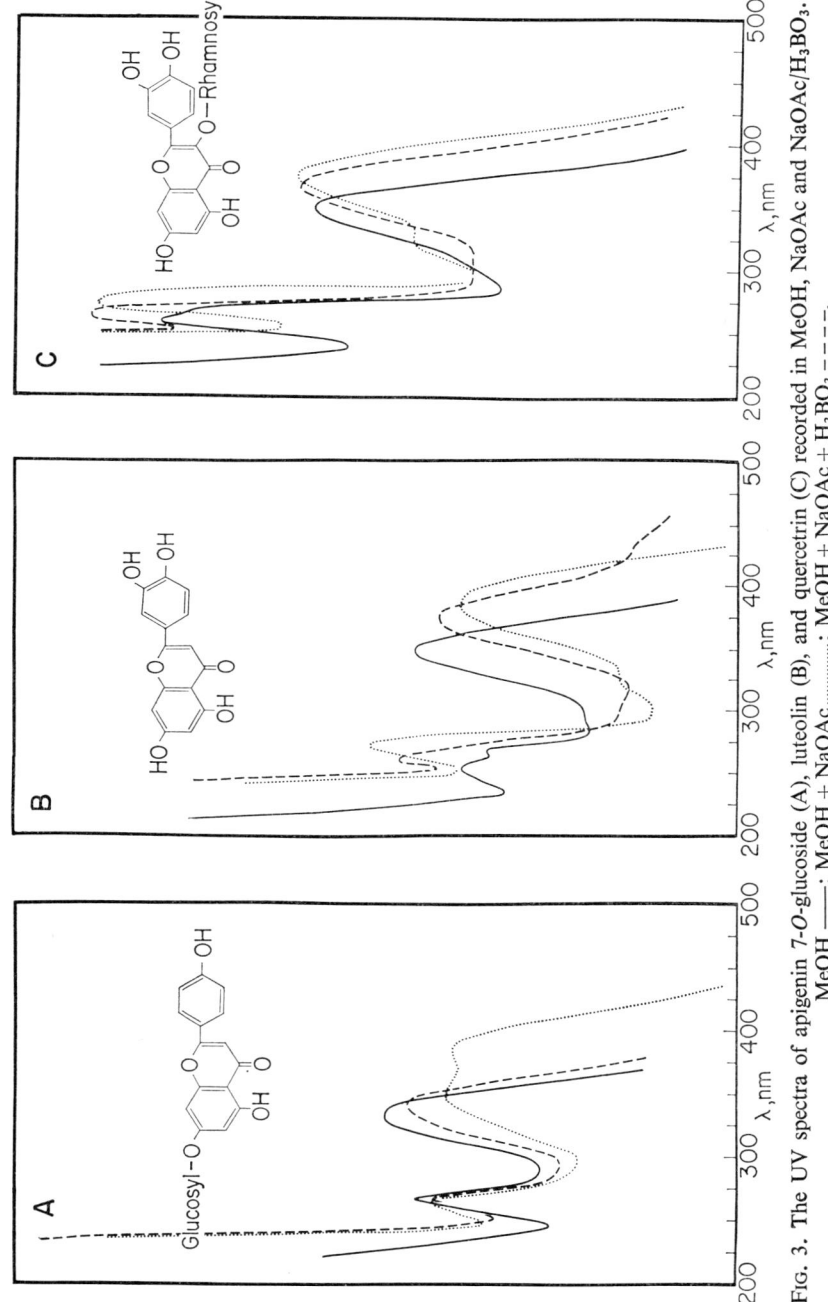

FIG. 3. The UV spectra of apigenin 7-O-glucoside (A), luteolin (B), and quercetin (C) recorded in MeOH, NaOAc and NaOAc/H$_3$BO$_3$. MeOH ———; MeOH + NaOAc ·········; MeOH + NaOAc + H$_3$BO$_3$ – – – –.

For this reason it is difficult to determine the presence or absence of free 7-hydroxyl groups in flavonoids possessing these oxygenation patterns unless the NaOAc spectrum is measured immediately after addition of the NaOAc to the cuvette.

7. The Detection of B-Ring Ortho-*dihydroxyl Groups in Flavones and Flavonols by the Effect of NaOAc/H₃BO₃ on the UV Spectrum*

In the presence of NaOAc, boric acid will chelate with *ortho*-dihydroxyl groups at all locations on the flavonoid nucleus, except at C-5,6. Such complexes are probably of type V.

(V)

Flavones and flavonols containing a B-ring *ortho*-dihydroxyl group show a consistent 12–30 nm bathochromic shift of Band I in the presence of NaOAc/ H₃BO₃ (Fig. 3).

8. The UV Spectra of Flavones and Flavonols in the Presence of AlCl₃ and AlCl₃/HCl

With aluminium chloride, flavones and flavonols which contain hydroxyl groups at C-3 or C-5 (Hörhammer *et al.* 1952; Jurd and Geissman, 1956) form acid stable complexes; in addition aluminium chloride forms acid labile complexes with flavonoids which contain *ortho*-dihydroxyl systems (Markham and Mabry, 1968). The structure of these complexes are probably of the types shown in Fig. 4. The complexes formed between AlCl₃ and the A- and B-ring *ortho*-dihydroxyl groups decompose in the presence of acid. In contrast, the AlCl₃ complex between the C-4 keto function and either the 3- or 5-hydroxyl group is stable in the presence of acid.

The 5-deoxyflavonol 3-*O*-glucoside, fisetin 3-*O*-glucoside, was hydrolyzed within a few minutes in the presence of anhydrous AlCl₃.

9. The Detection of Ortho-*dihydroxyl Groups in Flavones and Flavonols by the Effect of AlCl₃ and AlCl₃/HCl on the UV Spectrum*

Aluminium chloride has been used previously as a diagnostic reagent for the detection of *ortho*-dihydroxyl groups in anthocyanins (Geissman *et al.* 1953). In 1954, Harborne suggested the use of AlCl₃ for recognizing *ortho*-dihydroxyl groups in other flavonoids; however, the method was not fully

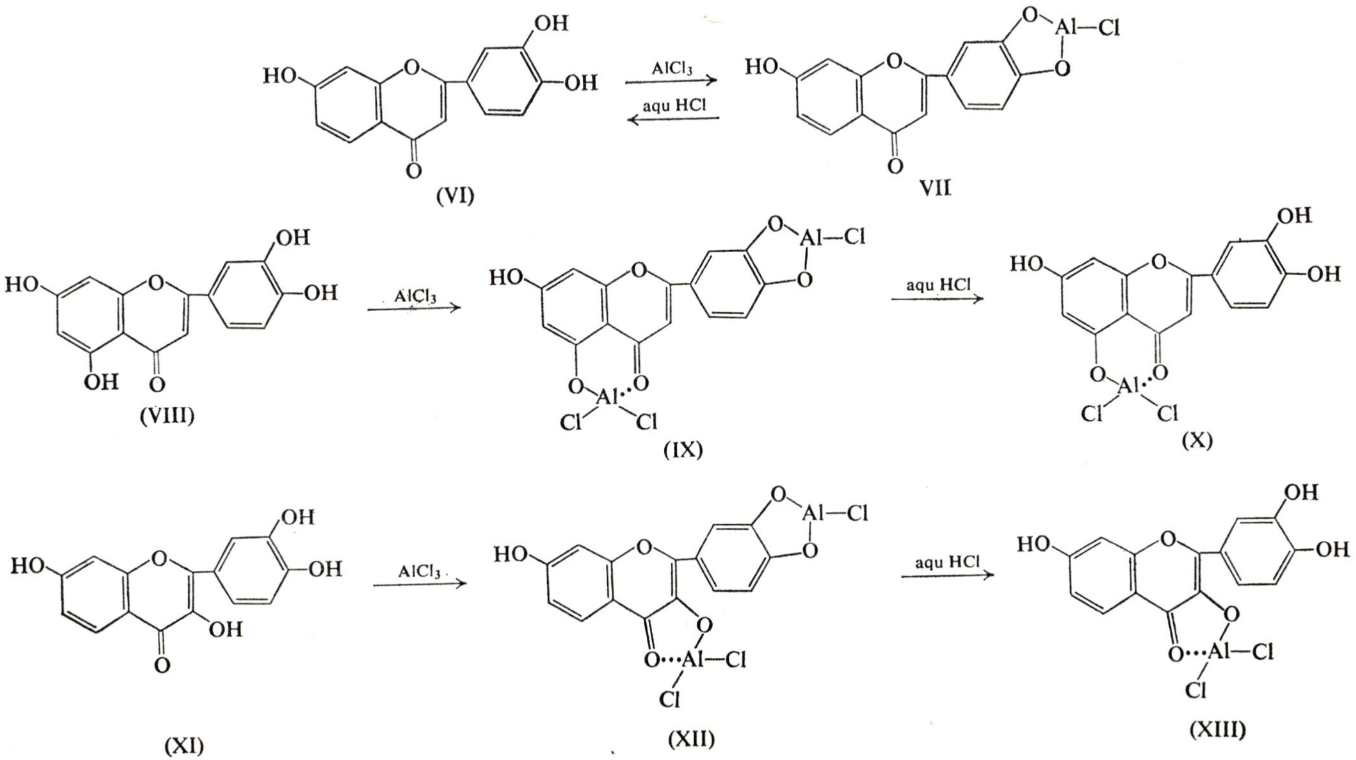

Fig. 4. Schemes illustrating the types of complexes that AlCl₃ can form with certain flavones and flavanols in the presence or absence of acid.

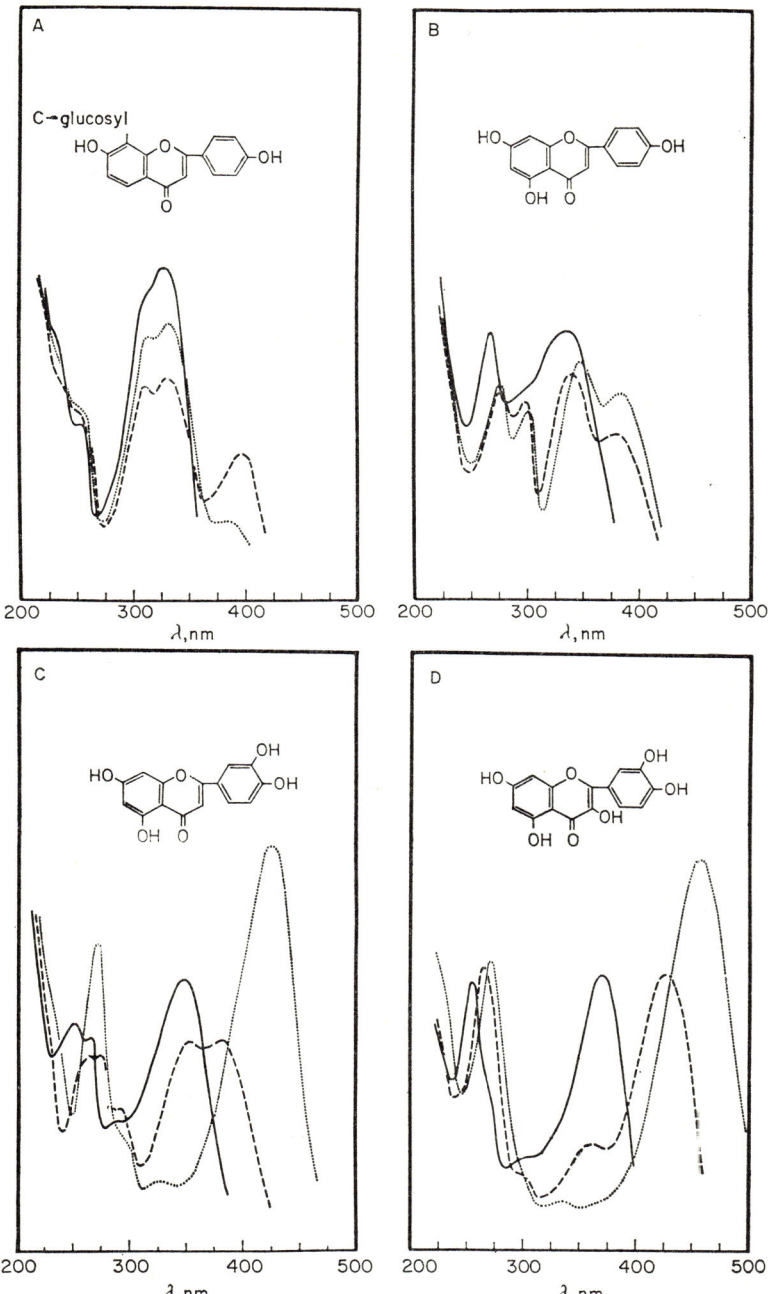

Fig. 5. The UV spectra of bayin (A), apigenin (B), luteolin (C) and quercetin (D) in MeOH, $AlCl_3$ and $AlCl_3/HCl$. MeOH ———; MeOH + $AlCl_3$ ....; MeOH + $AlCl_3$ + HCl – – – –.

developed by Harborne (1954), Swain (1954) or other investigators (Jurd and Geissman, 1956) at that time. The presence of an *ortho*-dihydroxyl group in the B-ring of flavones and flavonols can be detected by a comparison of the spectrum of the flavonoid in the presence of AlCl$_3$ and with that obtained in AlCl$_3$/HCl (Fig. 5). The hypsochromic shift (about 30–40 nm) observed in Band I (or Band Ia if Band I consists of two peaks) of the AlCl$_3$ spectrum on the addition of acid results from the decomposition of the complex of AlCl$_3$ with the *ortho*-dihydroxy group. The presence of three adjacent hydroxyl groups in the B-ring gives only a 20 nm hypsochromic shift on the addition of acid to the AlCl$_3$ solution.

10. The Detection of either 3- or 5-Hydroxyl Groups in Flavones and Flavonols by the Effect of AlCl$_3$/HCl on the UV Spectrum

The addition of acid to a methanolic solution of a flavone or flavonol which already contains AlCl$_3$ decomposes complexes between AlCl$_3$ and *ortho*-dihydroxyl groups; therefore any shift still remaining in Band I or Band II relative to the methanol spectrum will be due to AlCl$_3$ complexes with 3- and/or 5-hydroxyl groups in the flavonoid (Fig. 5). Regeneration of the methanol spectrum on the addition of acid indicates that both the 3- and 5-hydroxyl groups are either absent or substituted. The only difficulty encountered in the interpretation of the AlCl$_3$/HCl spectra was with members of the rare group of 5-deoxy-7-hydroxyflavones. With these compounds the spectrum observed after the addition of HCl to the solution used for the AlCl$_3$ spectrum exhibited all the peaks of the methanol spectrum but, in addition, showed a moderately intense peak about 60 nm from Band I of the methanol spectrum (Fig. 5A).

The AlCl$_3$/HCl spectrum of a 5-hydroxyflavone typically consists of four major absorption peaks, Bands Ia, Ib, IIa and IIb, which are all shifted bathochromically relative to their Band of origin (presumably Ia and Ib originate from I; and IIa and IIb from II) in the methanol spectrum (see spectra for apigenin, Fig. 5B).

There is a clear distinction between the magnitude of the AlCl$_3$/HCl bathochromic shifts associated with 5-hydroxyflavones and those observed for 3-hydroxyflavones. The bathochromic shifts of Band I (in MeOH) to Band Ia (in AlCl$_3$/HCl) in the spectra of 5-hydroxyflavones are in the range of 40–55 nm. In contrast, the shift is consistently around 60 nm for 3-hydroxyflavones.

B. THE UV SPECTRA OF ISOFLAVONES, FLAVANONES AND DIHYDROFLAVONOLS IN METHANOL

Isoflavones (XIV), flavanones (XV), and dihydroflavonols (XVI) all give similar UV spectra as a result of their having little or no conjugation between the B-ring and the C-4 carbonyl function. They are all readily distinguished from flavones and flavonols by their UV spectra, which typically exhibit an

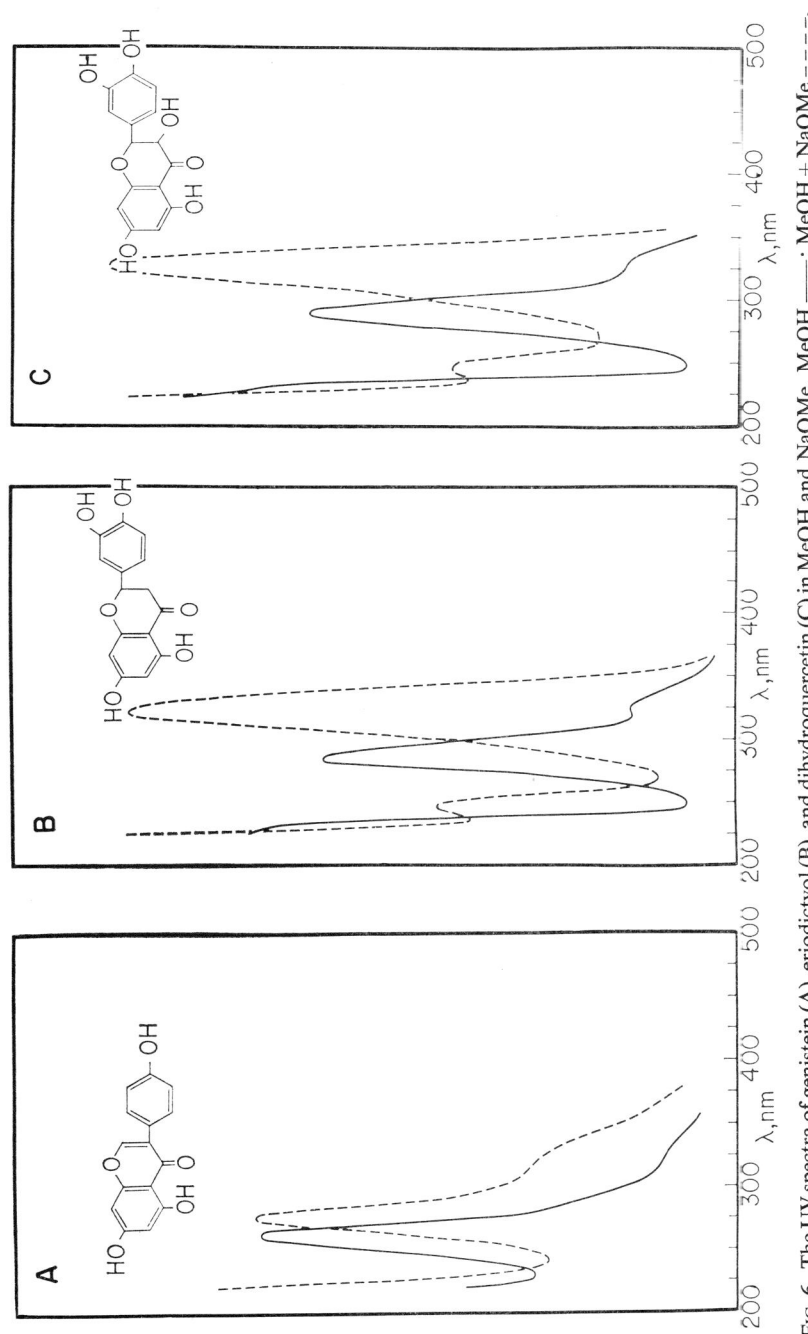

FIG. 6. The UV spectra of genistein (A), eriodictyol (B), and dihydroquercetin (C) in MeOH and NaOMe. MeOH ———; MeOH + NaOMe – – – –.

(XIV) Isoflavone skeleton   (XV) Flavanone skeleton

(XVI) Dihydroflavonol skeleton

intense Band II absorption with only a shoulder or low intensity peak representing Band I (Fig. 6).

The spectral data for a selected group of isoflavones, flavanones and dihydroflavanols are presented in Table VI.

1. The UV Spectra of Isoflavones in MeOH

The Band II absorption of isoflavones usually occurs in the region 245–270 nm, and is relatively unaffected by increased hydroxylation of the B-ring, cf. the Band II position in: 5,7-dihydroxyisoflavone (259 nm); 4′,5,7-trihydroxyisoflavone (261 nm) and 3′,4′,5,7-tetrahydroxyisoflavone (262 nm). Band II is, however, shifted bathochromically by increased oxygenation in the A-ring (Table V).

TABLE V

Band II in the UV spectra of isoflavones differing in their A-ring oxidation pattern

Isoflavone	Oxidation pattern		Band II (nm)
	A-ring	*B-ring*	
Daidzein	7	4′	249
Genistein	5,7	4′	261
6-Hydroxygenistein	5,6,7	4′	270

Some specific effects of isoflavone oxidation and substitution pattern on the UV spectra are:

     a. Simple polyoxygenated isoflavones which have their Band II absorption in the 260–270 nm range are usually trioxygenated in their A-ring.

     b. The UV spectra of 6,7-dioxygenated isoflavones, such as texasin and afrormosin are unusual (Dyke *et al.*, 1964; Markham *et al.* 1967) in that

TABLE VI

The ultraviolet spectra[a] of isoflavones, flavanones and dihydroflavonols[a]

Flavonoid	MeOH λ_{max} (nm) R.A.[a]		NaOMe λ_{max} (nm) R.A.		AlCl$_3$ λ_{max} (nm) R.A.		AlCl$_3$/HCl λ_{max} (nm) R.A.		NaOAc λ_{max} (nm) R.A.		NaOAc/ H$_3$BO$_3$ λ_{max} (nm) R.A.	
1. Daidzein (4′,7-dihydroxyisoflavone)	303sh	1·0	328	1·0	300	1·0	302	1·0	331	1·0	303	1·0
	259sh	2·5	289sh	1·4	260sh	2·3	263sh	2·2	310	1·0	261sh	2·1
	249	2·7	259	2·2	249	2·5	249	2·5	254	3·2		
2. Formononetin (4′-methoxy-7-hydroxyisoflavone	301	1·0	335	1·0	301	1·0	302	1·0	335	1·0	303	1·0
	258sh	2·6	255	2·8	261sh	2·4	261sh	2·4	313sh	1·0	264sh	2·1
	248	2·8			248	2·8	248	2·7	254	2·5		
3. Genistein (4′,5,7-trihydroxyisoflavone)	328sh	1·0	327sh	1·0	374	1·0	373	1·0	326	1·0	336sh	1·0
	261	6·4	275	2·5	307sh	1·6	308sh	1·5	271	3·6	262	6·5
					272	6·6	273	6·6				
4. Genistin (genistein 7-O-glucoside)	330sh	1·0	356sh	1·0	375	1·0	374	1·0	332sh	1·0	328sh	1·0
	261	12·2	271	8·7	309sh	1·8	272	9·7	261	6·3	261	9·6
					272	10·2						
5. Prunetin (genistein 7-methyl ether)	328sh	1·0	354sh	1·0	373	1·0	370	1·0	330sh	1·0	333sh	1·0
	262	13·2	272	9·5	309sh	1·8	310sh	1·5	262	8·4	262	9·0
					273	9·1	274	8·9				
6. Biochanin A (genistein 4′-methyl ether)	330sh	1·0	329	1·0	375	1·0	373	1·0	328	1·0	330sh	1·0
	261	8·9	273	3·8	310sh	1·6	310sh	1·6	272	3·2	262	4·9
					273	7·6	273	7·6				

TABLE VI (continued)

Flavonoid	MeOH λ_{max} (nm)	R.A.[a]	NaOMe λ_{max} (nm)	R.A.	AlCl$_3$ λ_{max} (nm)	R.A.	AlCl$_3$/HCl λ_{max} (nm)	R.A.	NaOAc λ_{max} (nm)	R.A.	NaOAc/ H$_3$BO$_3$ λ_{max} (nm)	R.A.
7. Texasin (4′-methoxy-6,7-dihydroxy-isoflavone)	325 255	1·0 2·5	352 254	1·0 2·1	344 252	1·0 2·2	325 257	1·0 2·6	339 253sh	1·0 2·4	328	1·0
8. Texasin 7-O-glucoside	324 258	1·0 3·8	367 254	1·0 4·2	324 258	1·0 3·5	323 258	1·0 3·5	372sh 330 256	1·0 1·2 5·2	325 258	1·0 3·6
9. Afrormosin (texasin 6′-methyl ether)	315 256	1·0 2·4	347 255	1·0 1·8	315 255	1·0 2·3	315 255	1·0 2·3	345 252sh	1·0 1·8	319 253sh	1·0 2·2
10. Orobol (3′,4′,5,7-tetrahydroxy-isoflavone)	337sh 294sh 262	1·0 2·2 5·6	(dec.) 334 269	1·0 1·0	367 298sh 270	1·0 4·0 5·8	371 272	1·0 7·2	318 270	1·0 2·4	293sh 265	1·0 1·5
11. Orobol 7-O-glucoside	322sh 290sh 262	1·0 3·1 8·2	(dec.) 337	1·0	367 297sh 269	1·0 3·8 5·4	375 272	1·0 6·7	330sh 261	1·0 5·0	259	1·0
12. Pratensein (orobol 4′-methyl ether)	292sh 263	1·0 2·6	322 270	1·0 2·6	372 311sh 273	1·0 1·8 7·9	371 273	1·0 8·9	325sh 271	1·0 3·0	296sh 263	1·0 2·4
13. Irigenin (3′,4′,6-trimethoxy-5,5′,7-trihydroxyisoflavone	337sh 267	1·0 5·1	336 273	1·0 2·1	372 316 276	1·0 2·6 7·3	374 316sh 278	1·0 2·4 8·4	338 273	1·0 2·4	339sh 268	1·0 4·7

Compound	λ (R.A.)	λ (R.A.)	λ (R.A.)	λ (R.A.)	λ (R.A.)	λ (R.A.)
14. Pinocembrin (5,7-dihydroxyflavanone)	326sh (1·0); 289 (3·0)	324 (1·0); 244 (0·2)	376 (1·0); 311 (6·7)	373 (1·0); 309 (5·8)	324 (1·0)	327sh (1·0); 291 (3·2)
15. Naringenin (4′,5,7-trihydroxyflavanone)	326sh (1·0); 288 (3·2)	323 (1·0); 245 (0·7)	375 (1·0); 312 (7·6)	372 (1·0); 310 (6·8)	323 (1·0)	332sh (1·0); 290 (3·4)
16. Sakuranin (4′,5-dihydroxy-7-methoxyflavanone 5-O-glucoside)	317sh (1·0); 280 (3·7)	393 (1·0); 316 (0·4)	312sh (1·0); 279 (2·6)	310sh (1·0); 280 (2·2)	314sh (1·0); 279 (3·0)	313sh (1·0); 278 (2·8)
17. Eriodictyol (3′,4′,5,7-tetrahydroxy-flavanone)	324sh (1·0); 288 (2·6)	324 (1·0); 246 (0·6)	378 (1·0); 311 (5·9)	374 (1·0); 309 (5·5)	326 (1·0); 290sh (0·4)	333sh (1·0); 290 (3·3)
18. Hesperidin (3′,5,7-trihydroxy-4′-methoxyflavanone 7-O-rutinoside)	326 (1·0); 283 (7·0)	356 (1·0); 286 (2·6); 242 (3·3)	383 (1·0); 308 (4·8)	379 (1·0); 306 (4·2)	329 (1·0); 284 (2·8)	327 (1·0); 284 (4·2)
19. Dihydrofisetin (3,3′,4,7-tetrahydroxy-flavanone)	310 (1·0); 277 (2·1)	(dec.) 333 (1·0); 252 (0·7)	351 (1·0); 305 (2·7)	308sh (1·0); 277 (2·1)	335 (1·0); 285 (0·5); 256sh (0·6)	314sh (1·0); 281 (1·8)
20. Taxifolin (dihydroquercetin; 3,3′,4′,5,7-pentahydroxyflavanone)	328sh (1·0); 289 (3·9)	(dec.) 328sh (1·0); 247 (0·5)	379 (1·0); 313 (4·6)	375 (1·0); 312 (4·9)	327 (1·0); 289sh (0·3)	337sh (1·0); 292 (3·1)
21. Dihydrorobinetin (3,3′,4′,5′,7-pentahydroxyflavanone)	308 (1·0); 275 (2·0)	(dec.) 334 (1·0); 251 (0·5)	346sh (1·0); 307 (2·5); 277sh (1·6)	307 (1·0); 275 (2·0)	334 (1·0); 280 (0·5); 257sh (0·7)	313sh (1·0); 278 (1·6)

[a] The relative absorptivities (R.A.) are presented for the λ_{max}'s obtained for each spectrum using the longest wavelength peak or shoulder as 1·0.

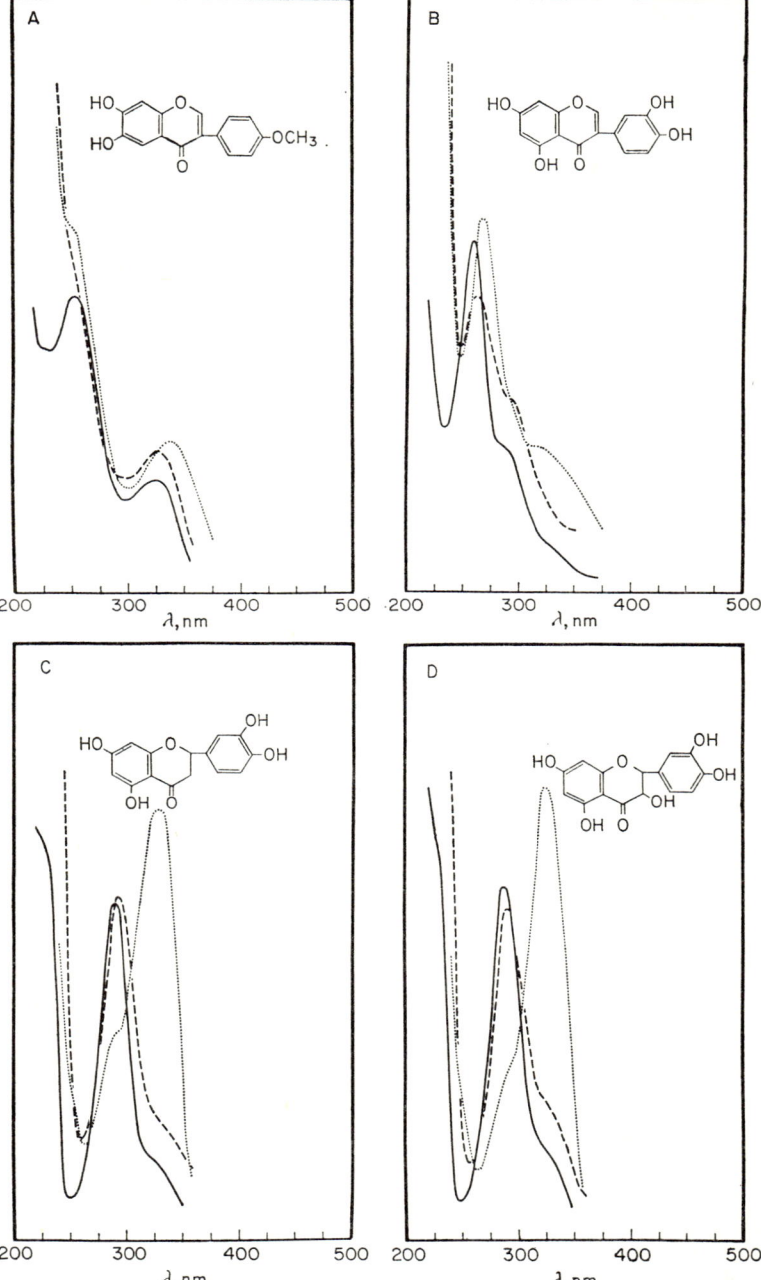

FIG. 7. The UV spectra of texasin (A), orobol (B), eriodictyol (C), and dihydroquercetin (D) in MeOH——; MeOH + NaOAc ·········; MeOH + NaOAc/H₃BO₃- - - -.

Band I is abnormally intense (see Fig. 7A); thus the spectra appear similar to those observed for flavones (however, paper chromatography or NMR spectroscopy clearly identifies them as isoflavones).

c. Methylation or glycosylation of either 7- or 4'-hydroxyl groups in isoflavones has little or no effect on the UV spectrum while substitution of 5-hydroxyl groups causes a 5–10 nm hypsochromic shift of Band II.

d. Loss of a 5-hydroxyl group causes a 9–15 nm hypsochromic shift in Band II.

2. The UV Spectra of Flavanones and Dihydroflavonols in MeOH

The UV spectra of dihydroflavonols are almost identical with those obtained for the equivalent flavanones (Fig. 6); thus the presence or absence of the C-3 hydroxyl group in flavonoids which do not have a C_2-C_3 double bond makes little difference to the UV spectra. Both flavanones and dihydroflavonols have their major absorption peak (Band II) in the range 270–295 nm and are therefore clearly distinguished from the spectra of isoflavones (which have their Band II peaks between 245 and 270 nm). Removal of the 5-hydroxyl group from a flavanone or dihydroflavonol causes a 10–15 hypsochromic shift of the major absorption band, i.e., Band II. Increasing oxygenation in the B-ring of flavanones and dihydroflavonols has no noticeable effect of their UV spectra.

3. The UV Spectra of Isoflavones, Flavanones and Dihydroflavonols in the Presence of NaOMe

The spectra of isoflavones containing A-ring hydroxyl groups usually show bathochromic shifts of both Band I and Band II in the presence of NaOMe (Fig. 6). In addition, the peaks in the UV spectra of a solution of 3',4'-dihydroxyisoflavones with added NaOMe show reduced intensity within a few minutes (Fig. 8). The only other isoflavone showing signs of decomposition in NaOMe was 6-hydroxygenistein (4',5,6,7-tetrahydroxyisoflavone), which contains the alkali sensitive 5,6,7-hydroxylation pattern.

The UV spectra of all flavanones and dihydroflavonols with A-ring hydroxylation show bathochromic shifts for Band II with NaOMe (Fig. 6). For dihydroflavonols, the magnitude of the bathochromic shift depends upon the presence or absence of a free 5-hydroxyl group. Spectra of 5,7-dihydroxy-dihydroflavonols exhibit a consistent 35–40 nm shift of the major absorption peak, whereas spectra of the 7-hydroxy-dihydroflavonols lacking a free 5-hydroxyl group show a 55–60 nm shift. In both cases, an increase in the Band II peak intensity is observed.

The UV spectra of flavanones in the presence of NaOMe also exhibit bathochromic shifts of the main absorption band (Band II) of about 35 nm for 5,7-dihydroxyflavanones (Fig. 6) and 60 nm for 7-hydroxyflavanones. Again, the shifts are accompanied by an increase in the intensity of Band II. Under alkaline conditions, however, some flavanones (in particular those lacking a free 5-hydroxy group, Jurd and Horowitz, 1961; Narasimhachari and Seshadri,

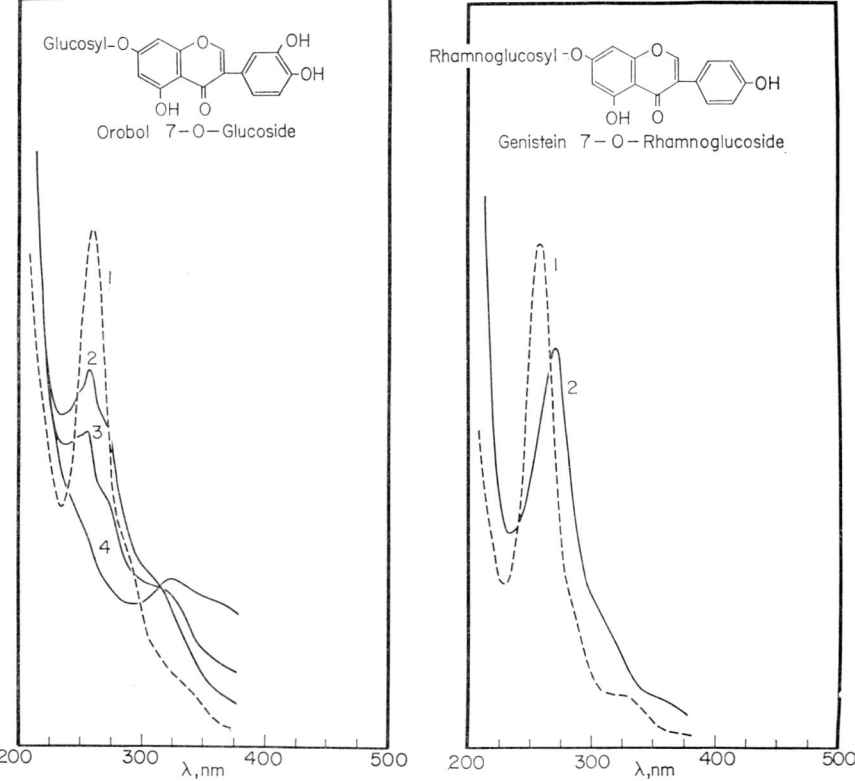

Fig. 8. The different effects of NaOMe on the spectra of isoflavones having a 3′,4′-dihydroxyl group (orobol 7-O-glucoside; 1, in MeOH; 2, NaOMe immediately; 3, after 5 min; 4, after 10 min) and a 4′- hydroxyl group (genistein 7-O-rhamnoglucoside; 1, in MeOH; 2, NaOMe immediately and after 20 min).

1948, 1951) will isomerize to chalcones, which have an entirely different UV spectrum in the presence of NaOMe (see Fig. 9).

Flavanones with 5,6,7 or 6,7,8 hydroxylation patterns decompose in the presence of NaOMe and, as a consequence, their UV spectra degenerate.

4. The UV Spectra of Isoflavones, Flavanones and Dihydroflavonols in the Presence of NaOAc

NaOAc specifically ionizes the 7-hydroxyl group in isoflavones (Jurd and Horowitz, 1961). Unlike many flavones and flavonols, isoflavones do not contain ionizable 3- or 4′-hydroxyl groups, and therefore the interpretation of the spectral shifts with NaOAc is simplified. NaOAc causes Band II of the UV spectrum of a 7-hydroxyisoflavone to shift 6–20 nm bathochromically; however, as was previously observed for flavones little or no shift occurs when there is an oxygen substituent at position 6.

The presence or absence of a free 7-hydroxyl group in flavanones and

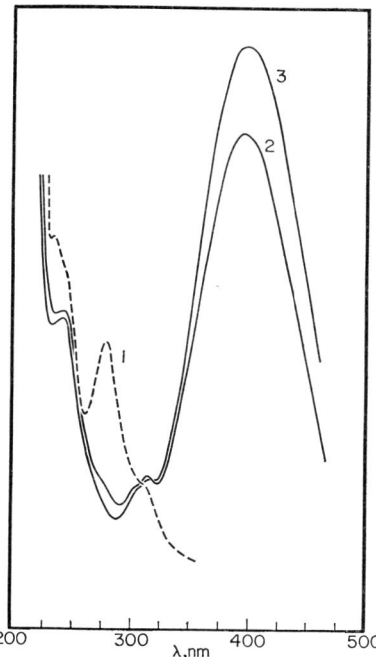

Sakuranin

FIG. 9. The UV spectrum of sakuranin in the presence of NaOMe 1, immediately; 2, after 3 min; 3, after 8 min; the spectra illustrate the conversion of sakuranin to the equivalent ionized chalcone.

dihydroflavonols may readily be determined from their UV spectra by comparing the positions of the major absorption peak (Band II) in the methanol spectrum with that of the same peak in the NaOAc spectrum (Fig. 7). The shift is about 35 nm for 5,7-dihydroxyflavanones and 5,7-dihydroflavanols and about 60 nm for their 5-deoxy-equivalents. Degeneration of the UV spectrum with time was obesrved for 5,6,7-trihydroxyflavanone.

5. *The UV Spectra of Isoflavones, Flavanones and Dihydroflavonols in the Presence of AlCl₃ and AlCl₃/HCl*

Band II in the UV spectra of 5-hydroxyisoflavones undergoes a consistent 9–13 nm bathochromic shift (relative to the spectrum in methanol) in the presence of AlCl₃/HCl. With 5-hydroxyflavanones and 5-hydroxydihydroflavanols

2*

TOM J. MABRY

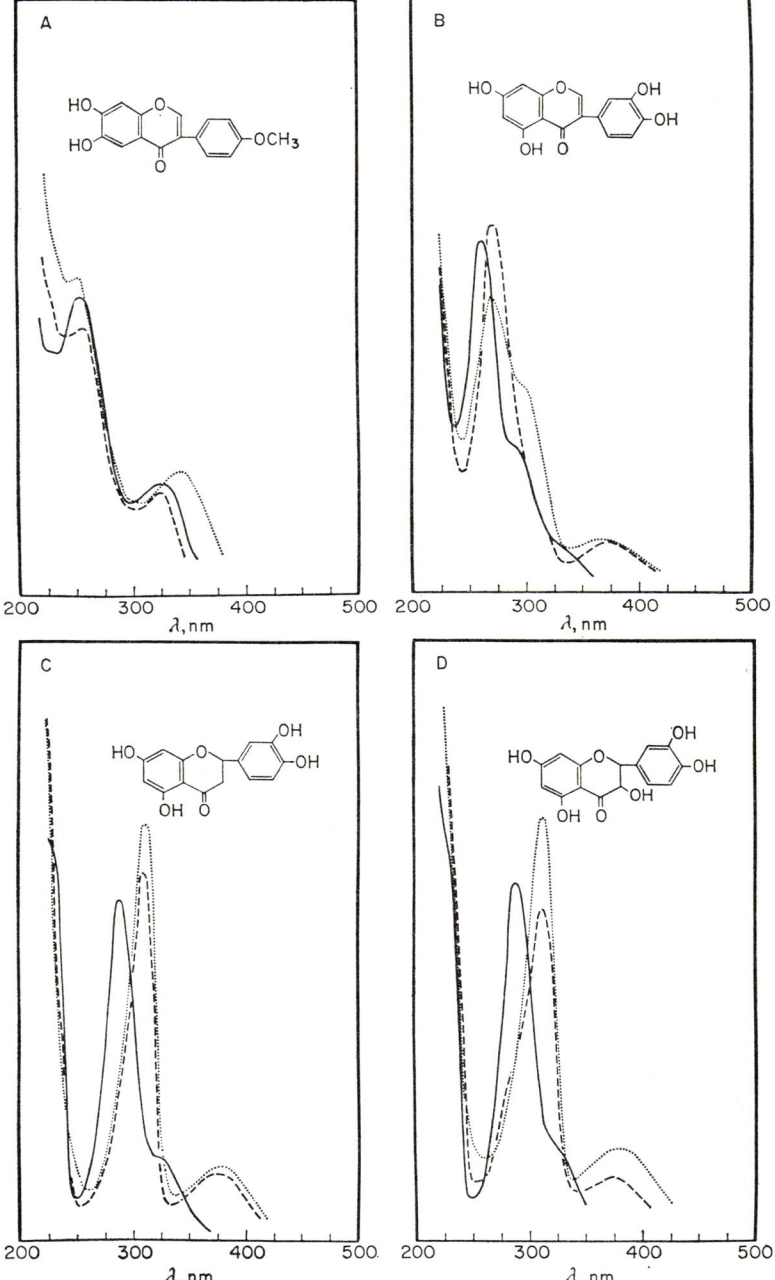

FIG. 10. The UV spectra of texasin (A), orobol (B), eriodictyol (C) and dihydroquercetin (D) in MeOH, AlCl₃ and AlCl₃/HCl. MeOH ———; MeOH + AlCl₃ ·············; MeOH + AlCl₃ + HCl – – – –.

the $AlCl_3/HCl$ reagent also causes a consistent bathochromic shift (22–26 nm) of Band II (Fig. 10).

III. FLAVONOID STRUCTURE ANALYSIS BY NUCLEAR MAGNETIC RESONANCE SPECTROSCOPY

A. INTRODUCTION

The application of nuclear magnetic resonance (NMR) spectroscopy to the structure analysis of flavonoids is now well established. Many flavonoid aglycones, in particular isoflavones and highly methylated flavones and flavonols, are sufficiently soluble in the commonly used solvent, deuterio-chloroform ($CDCl_3$) for direct NMR analysis. However, most naturally occurring flavonoids, including all of the flavonoid glycosides, have low solubility in $CDCl_3$; therefore, prior to 1964 most workers were limited to the NMR analysis of the more soluble, methyl, ethyl and acetyl derivatives (see e.g., Massicot and Marthe, 1962; Massicot *et al.* 1963). However, the signals observed for the substituent groups in these derivatives often obscure signals of other protons in the flavonoid.

In 1964–65, two groups of workers (Waiss *et al.* 1967 and Mabry *et al.* 1965); independently investigated the usefulness of trimethylsilyl ether derivatives for obtaining NMR spectra of flavonoids which were otherwise insoluble in $CDCl_3$. At about the same time (Batterham and Highet, 1964) hexa-deuteriodimethylsulphoxide (DMSO-d_6) was introduced as a solvent for the direct NMR analysis of flavonoids.

B. THE USE OF DMSO-d_6 AS A SOLVENT FOR FLAVONOID NMR SPECTROSCOPY

DMSO-d_6 has been used extensively as a solvent for investigations of flavonoid structures by NMR spectroscopy (Batterham and Highet, 1964; Grouiller, 1966; Grouiller and Pacheco, 1967). Some of the advantages of this method relative to other available procedures include those listed below.

a. Most flavonoid aglycones and glycosides are sufficiently soluble in DMSO-d_6 for direct NMR analysis, thereby eliminating the necessity of preparing derivatives.

b. The proton signals resulting from the CH_3 groups in the small amount of DMSO always present in DMSO-d_6 occur in a narrow band between 2·4 and 2·6 ppm (δ-scale) outside the region where most flavonoid protons absorb.

c. DMSO-d_6 (if it is anhydrous) can be used specifically for observing protons on phenolic hydroxyl groups. For example, in the flavonoid aglycone galangin (3,5,7-trihydroxyflavone), the hydroxyl proton signals are readily distinguishable (Fig. 11). Traces of water in the DMSO-d_6, however, cause the flavonoid hydroxyl proton signals to broaden.

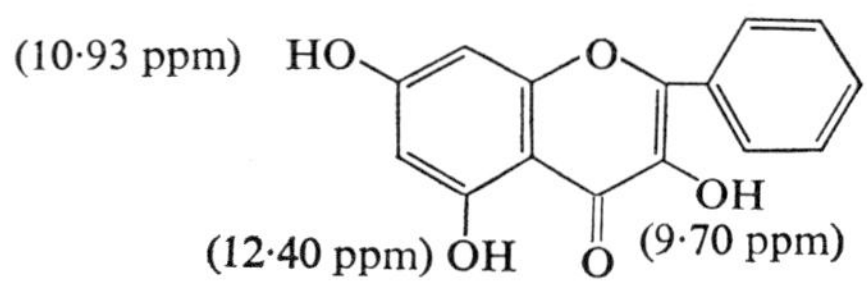

FIG. 11. Chemical shifts (δ) as observed in DMSO-d_6 for hydroxyl protons in galangin.

There are, however, also a number of disadvantages associated with the use of DMSO-d_6 as solvent for flavonoid spectroscopy.

a. DMSO-d_6 has a boiling point of 189°, which makes recovery of the flavonoid difficult.

b. DMSO-d_6 rapidly absorbs atmospheric moisture, and the signal obtained from the adsorbed H_2O [variable around $3 \cdot 5$ ppm (δ)] often obscures NMR signals resulting from the flavonoid protons.

c. DMSO-d_6 must be handled carefully since it rapidly penetrates the skin carrying with it any dissolved substances.

d. Some flavonoids have been reported (Grouiller and Pacheco, 1967) to undergo decomposition in DMSO-d_6.

C. PREPARATION OF TRIMETHYLSILYL ETHER DERIVATIVES OF FLAVONOIDS

Waiss *et al.* 1967 and Mabry *et al.* 1965 demonstrated the usefulness of trimethylsilyl ether derivatives for the NMR analysis of flavonoids. The method offers several advantages.

a. Most flavonoids can be converted to trimethylsilyl ether derivatives and prepared for NMR analysis in about 20 minutes.

b. Trimethylsilyl ether derivatives of flavonoids are miscible with carbon tetrachloride (CCl_4) in all proportions.

c. The signals of the trimethylsilyl protons occur at fields higher than $0 \cdot 5$ ppm (δ), and are therefore well out of the absorption region of flavonoid protons.

d. Both the preparation and the hydrolysis of trimethylsilyl ether derivatives are quantitative and may be carried out under mild conditions (see Fig. 12).

Standard procedures for the preparation of TMS ethers of flavonoids as well as some 130 spectra (with interpretations) are presented elsewhere by Mabry *et al.* (1969).

D. INTERPRETATION OF THE NMR SPECTRA OF FULLY TRIMETHYL-SILYLATED FLAVONOIDS

Proton signals obtained in the NMR spectra of trimethylsilylated flavonoids generally occur in the range 0–8 ppm, δ-scale (Fig. 13). All the NMR data

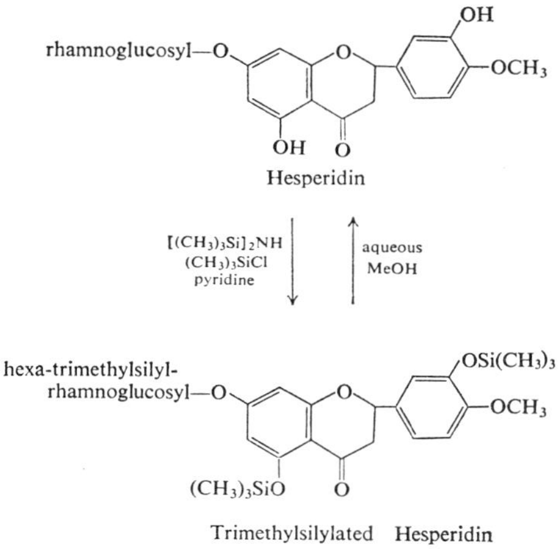

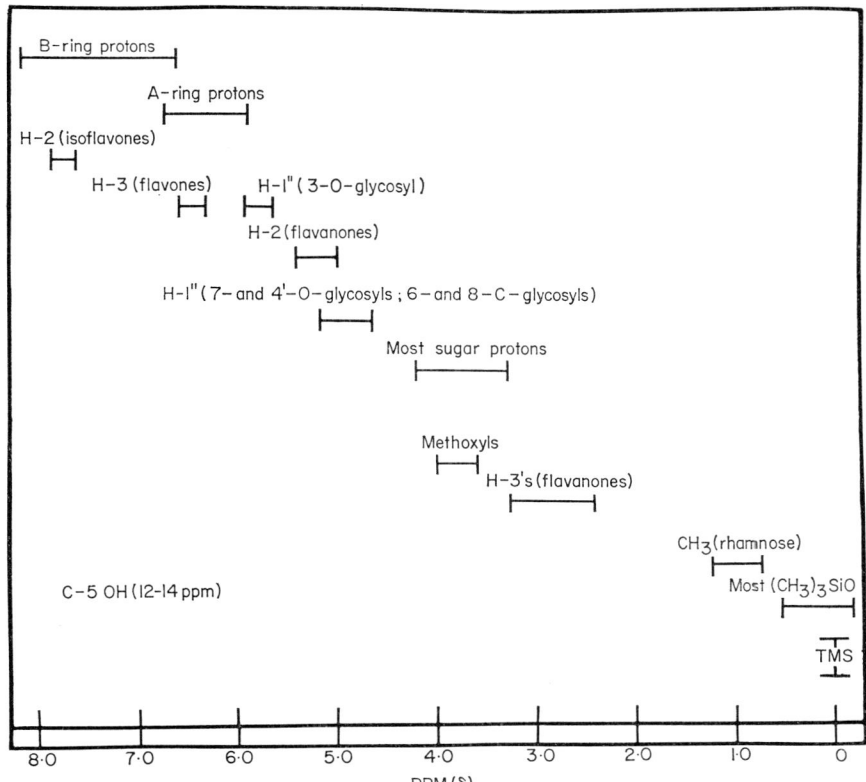

FIG. 12. The reaction scheme for trimethylsilylation of hesperidin and the recovery of the flavonoid by hydrolysis of the derivative.

FIG. 13. Approximate chemical shift ranges for the protons in trimethylsilylated flavonoids.

presented in this section are for the TMS ethers of flavonoids and are recorded in the δ-scale. All the NMR data discussed here was determined with a Varian A-60 spectrometer.

The numbering system used in the NMR discussions for flavonoids is as given in (I) (see p. 2).

1. A-Ring Protons

a. C-6 and C-8 protons. The protons at C-6 and C-8 of flavones, flavonols and isoflavones which contain the common 5,7-dihydroxy substitution pattern give rise to two doublets ($J = 2.5$ c.p.s.) in the range 6.0–6.5 ppm. The H-6 doublet occurs consistently at higher field than the signal for H-8 (Table VII).

TABLE VII

Chemical shifts of C-6 and C-8 protons in 5,7-dihydroxyflavonoids and their glycosides

Type of flavonoid	H-6[a] (δ, ppm)	H-8[a] (δ, ppm)
Flavones, Flavonols, Isoflavones		
	6.0–6.2	6.3–6.5
	6.2–6.4	6.5–6.9
Flavanones, Dihydroflavonols		
	5.75–5.95	5.9–6.1
	5.9–6.1	6.1–6.4

[a] Doublet, $J = 2.5$ c.p.s.

When a sugar is attached to the oxygen at C-7 the signals for both H-8 and H-6 are shifted downfield (Table VII).

In flavanones and dihydroflavonols which contain the 5,7-dihydroxy substitition the signals for the A-ring protons appear at higher field than in the corresponding flavones and flavonols (Table VII).

  b. *Distinguishing between C-6, C-8, and C-3 proton signals.* The only other proton of the flavonoid nucleus which gives a signal consistently in the same region of the NMR spectrum as those of the C-6 and C-8 proton is the C-3 proton of flavones, which appears as a singlet near 6·3 ppm. Many flavones have only one A-ring proton (e.g. 5,6,7- or 5,6,8-substitution patterns) and in these compounds the long A-ring proton produces a singlet which is often in the same region as the C-3 proton signal. A useful technique for distinguishing the C-6, C-8 and C-3 protons in such cases is to compare the NMR spectrum of the fully trimethylsilylated compound with that of the compound containing trimethylsilyl substituents on all hydroxyl groups but the one at C-5 (Mabry *et al.* 1965b; Seikel and Mabry, 1965; Hörhammer *et al.* 1965). In the spectrum of the partially trimethylsilylated compound (5-hydroxyl group unsubstituted) relative to the spectrum of the fully trimethylsilated flavonoid, the C-3 proton singlet is shifted downfield (usually 0·15 ppm or more), the C-8 proton is shifted upfield 0·15 ppm or more, while the C-6 proton is almost unaffected (shifted upfield 0·05 ppm or less). These shifts are clearly observed* when the spectrum of fully trimethylsilylated luteolin is compared with the spectrum of luteolin 3′,4′,7-trimethylsilyl ether (Fig. 14).

  c. *C-5, C-6 and C-8 protons in 7-hydroxyflavones, 7-hydroxyisoflavones, 7-hydroxyflavanones, 7-hydroxyflavanols and 7-hydroxydihydroflavonols.* Some naturally occurring flavonoids have only C-7 oxygenation in the A-ring. In these compounds the C-5 proton is deshielded by the C-4 keto group and therefore absorbs near 8·0 ppm (Table VIII), and thus at lower field than most aromatic protons. The C-5 proton in these 7-oxygenated flavonoids appears as a doublet ($J = ca$ 9 c.p.s.) as a result of *ortho* coupling between the C-5 and C-6 protons. The signals for the protons on C-6 and C-8 both occur at lower field than in the 5,7-dihydroxyflavonoids. Moreover, in 5-deoxyflavonoids the C-8 proton may absorb at either lower or higher field than the C-6 proton (in 5,7-dihydroxyflavonoids, as far as is known, the C-6 proton always absorbs at higher field than the C-8 proton).

2. B-Ring Protons

  a. *The C-2′, C-3′, C-5′ and C-6′ protons in 4′-oxygenated flavonoids.* The protons of ring B appear in the range 6·7–7·9 ppm, which is downfield from

---

* For some 6-C-glycosylflavones, the NMR spectrum of the fully trimethylsilylated flavonoid differs very little from the one obtained for the partially (C-5 OH free) trimethylsilylated compound with the exception that in the latter spectrum, a new signal near 12·5 ppm is observed. The reason for this is that the signals for the H-3 and H-8 protons are often separated by about 0·15 ppm in the spectrum of the fully trimethylsilylated flavonoid, thus a shift of about 0·15 ppm by both signals, one upfield and one downfield, produces a spectrum which appears unchanged.

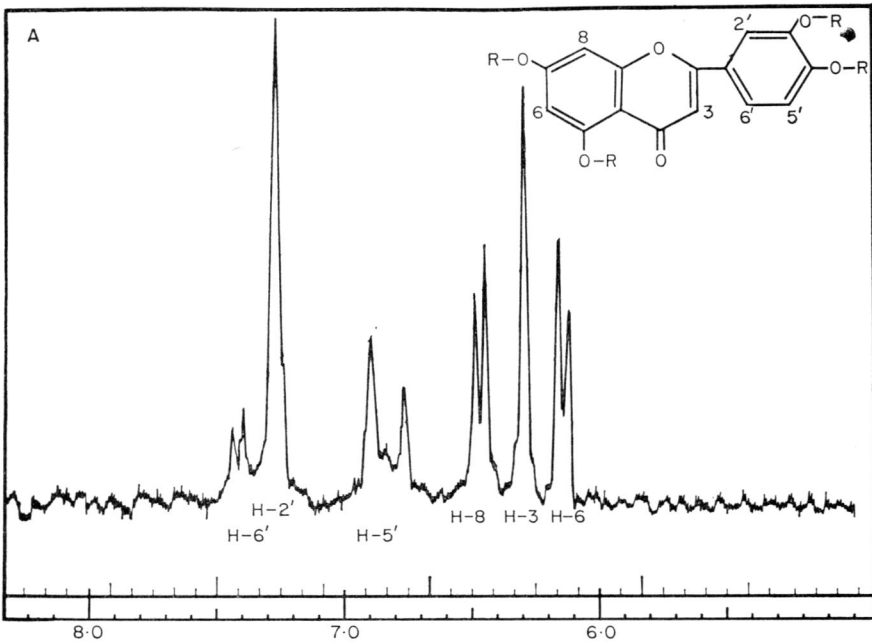

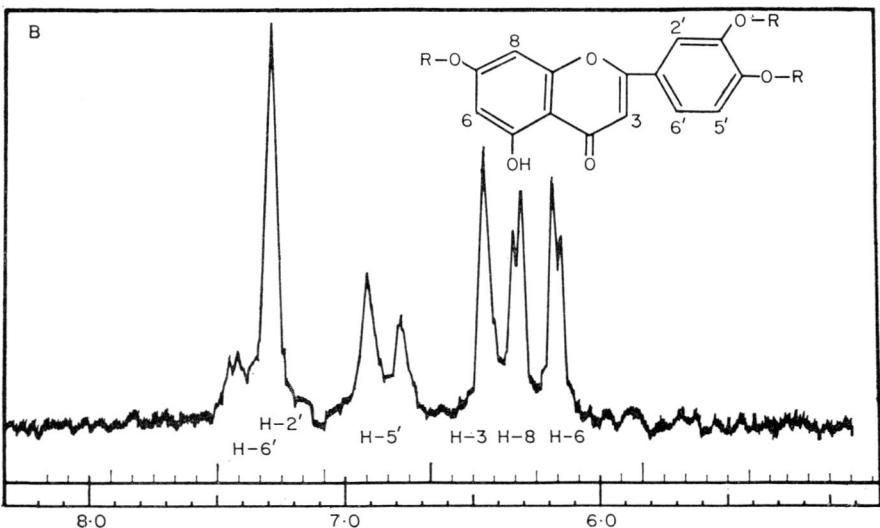

Fig. 14. (A) NMR spectrum of luteolin 3′,4′,5,7-tetra-trimethylsilyl ether (R = Si[CH₃]₃) and (B) luteolin 3′,4′,7-trimethylsilyl ether (R = Si[CH₃]₃) in CCl₄.

TABLE VIII

Chemical shifts of C-5, C-6 and C-8 protons in 7-oxygenated flavonoids

Type of flavonoid	H-5[a] (δ, ppm)	H-6[b] (δ, ppm)	H-8[c] (δ, ppm)
(flavanone structure: RO at C-7, positions 5, 6, 8 labelled; O, R', R'')	7·9–8·2	6·7–7·1	6·7–7·0
(flavone structure: RO at C-7, positions 5, 6, 8 labelled; O, R', R'')	7·7–7·9	6·4–6·5	6·3–6·4

[a] Doublet, $J = 9$ c.p.s.
[b] Quartet, $J = 9$ and 2·5 c.p.s.
[c] Doublet, $J = 2·5$ c.p.s.

the region where the A-ring protons usually absorb. The signal pattern observed for the B-ring protons is characteristic for the substitution pattern of that ring and, in addition, usually suggests the oxidation level of ring C.

If ring B is oxygenated only at C-4′, a typical four-peak pattern of two doublets (each $J = 8·5$ c.p.s.) is observed. The doublet for the C-3′ and -5′ protons, which are shielded by the C-4′ oxygen substituent, always appears upfield from the C-2′ and -6′ protons and generally falls in the range 6·65–7·1 ppm for all types of flavonoids. The position of the C-2′, -6′ doublet depends to some extent on the oxidation level of ring C (see Table IX); however, it consistently appears at lower field (7·1–8·1 ppm) than the C-3′, -5′ doublet.

TABLE IX

Chemical shifts of C-2′, -6′ and C-3′, -5′ protons in 4′-oxygenated flavonoids

4′-Oxygenated flavonoids	H-2′, -6′[a] (δ, ppm)
Flavanones	7·1–7·3
Dihydroflavonols	7·2–7·4
Isoflavones	7·2–7·5
Flavones	7·7–7·9
Flavonols	7·9–8·1

In all cases the H-3′, -5′ have chemical shifts of 6·5–7·1 δ, ppm.
[a] Doublet, J ca 8·5 c.p.s.

b. C-2′, C-5′ and C-6′ protons in 3′,4′-oxygenated flavonoids. The NMR spectra of flavonoids with the 3′,4′-oxygenation pattern in the B-ring are more complex than for compounds with the 4′-oxygenation.

The C-5′ proton of 3′,4′-oxygenated flavones and flavonols appears as a doublet centred between 6·7 and 7·1 ppm ($J = 8·5$ c.p.s.), and the C-2′ and -6′ proton signals, which often overlap, usually occur between 7·2 and 7·9 ppm. The relative positions of the signals for the C-2′ and -6′ protons may be used to distinguish the 3′-methoxy-4′-hydroxy from the 4′-methoxy-3′-hydroxy B-ring substitution pattern in flavones and flavonols. The C-2′ proton signal is usually centred at slightly higher field than the C-6′ proton signal in flavonoids containing a 4′-methoxyl group (Table X). In contrast, their positions are reversed when a 3′-methoxyl group is present in the 3′,4′-oxygenated compound.

Different spectral patterns are observed for 3′,4′-oxygenated isoflavones, flavanones and dihydroflavonols. These compounds give a complex multiplet, usually two peaks, for the C-2′, -5′ and -6′ protons in the region 6·7–7·1 ppm; the chemical shifts for these protons depend upon whether or not the proton is *ortho* or *para* to an oxygen function. In 3′,4′-dioxygenated isoflavones, flavanones and dihydroflavonols, the C-2′, -5′ and -6′ protons are either *ortho* or *para* to an oxygen substituent and therefore have similar chemical shifts.

c. C-2′ and C-6′ protons in 3′,4′,5′-oxygenated flavonoids. The C-2′ and -6′ proton signals in the spectra of 3′,4′,5′-oxygenated flavonoids usually overlap in the region 6·5–7·5 ppm in flavonoids having the 3′,4′,5′-oxygenation pattern.

3. C-Ring Protons

a. C-3 proton in flavones; C-2 protons in isoflavones. Considerable variation is found in the chemical shifts of the C-ring protons among the different flavonoid classes depending upon the oxidation level of the C-ring. The C-3 proton in flavones gives a sharp singlet near 6·3 ppm, in the region where the signals produced by the A-ring protons occur. On the other hand, the C-2 proton in isoflavones, which is in a beta position to the C-4 keto function, occurs in the range 7·6–8·7 ppm, a region downfield from where most A- and B-ring proton signals appear.

b. C-2 and C-3 protons of flavanones and dihydroflavonols. The signal for the C-2 proton of flavanones appears as a quartet (two doublets, $J_{cis} = ca$ 5 c.p.s., $J_{trans} = ca$ 11 c.p.s.) near 5·2 ppm as a result of the coupling of the C-2 proton with the two C-3 protons. The C-3 protons couple with each other ($J = 17$ c.p.s.) in addition to their spin-spin interaction with the C-2 proton, thus giving rise to two overlapping quartets centred around 2·8 ppm. Two of the signals of each quartet, however, are weak and are often not observed.

In dihydroflavonols the C-2 proton signal occurs as a doublet ($J = ca$ 11 c.p.s.) near 5·2 ppm, while the C-3 proton doublet appears further upfield at about 4·3 ppm. The 11 c.p.s. coupling constant is typical for 1,2-diaxial

Table X

Chemical shifts of C-2′ and C-6′ protons in 3′,4′-oxygenated flavones and flavonols

Type of flavonoid	H-2′[a] (δ, ppm)	H-6′[b] (δ, ppm)
(a) R = R′ = H (b) R = H, R′ = CH₃ (c) R = CH₃, R′ = H	7·2–7·3	7·3–7·5
R = H or CH₃ R′ = H or CH₃	7·5–7·7	7·6–7·9
	7·2–7·5	7·3–7·7
R = H, CH₃ or glucosyl	7·6–7·8	7·4–7·6

[a] Doublet, $J = 2·5$ c.p.s.
[b] Quartet, $J = 2·5$ and $8·5$ c.p.s.

protons; thus, the naturally occurring dihydroflavonols, which are known to have the R absolute configuration at both C-2 and C-3, can be represented by the structure shown in Fig. 15.

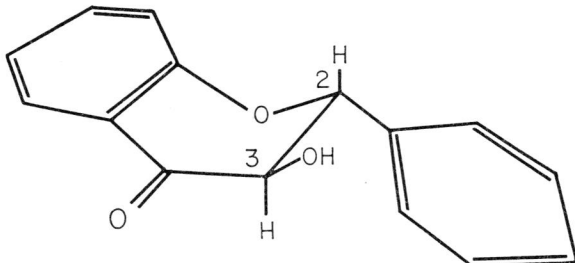

FIG. 15. The absolute stereochemistry (2R, 3R) of naturally occurring dihydroflavonols.

4. Sugar Protons

*a. The C-1″ Proton of flavonoid monosaccharides.** The chemical shift of the C-1″ proton of a sugar directly attached to the flavonoid hydroxyl group depends both on the nature of the flavonoid and on the position and stereochemistry of attachment to it.

With glucosides a sugar on the C-3 hydroxyl group can be readily distinguished from one at C-4′, C-5 or C-7 (see Table XI). In the latter three types of flavonoid O-glucosides, the C-1″ proton signal occurs near 5·0 ppm, while in flavonol 3-O-glucosides the C-1″ proton signal appears further downfield at about 5·8 ppm.

TABLE XI

Chemical shifts of the C-1″ protons of glucosyl and rhamnosyl substituents which are directly attached to flavonoid hydroxyl groups

Flavonoid glycoside	Sugar C-1″ proton (δ, ppm)
Flavonoid 7-O-glucoside	4·8–5·2
Flavonol 3-O-glucoside	5·7–5·9
Flavonol 7-O-rhamnoside	5·1–5·3
Flavonol 3-O-rhamnoside	5·0–5·1
Dihydroflavonol 3-O-glucoside	4·1–4·3
Dihydroflavonol 3-O-rhamnoside	4·0–4·2

Glucose commonly forms a β-linkage in flavonoid glycosides and the C-1″ proton of the β-linked sugar has a diaxial coupling with the C-2″ proton. Thus the C-1″ proton usually appears as a doublet with a coupling constant of about 7 c.p.s. In flavonoid 7-O-glucosides, however, the glucosyl C-1″ proton does not appear as a sharp doublet but instead gives a complex multiplet. Dreiding models indicate that the reason for this is that the 7-O-glucosyl moiety rotates

---

* In flavonoid glycosides, the protons in the sugar attached directly to the flavonoid are denoted as C-1″, C-2″ and so on; in disaccharides the protons of the terminal sugar are designated C-1‴, C-2‴ and so on.

with respect to the flavonoid nucleus and thus the C-1″ proton of the sugar experiences different electronic environments. As rotation becomes more restricted by C-8 and/or C-6 substituents, the glucosyl C-1″ proton signal tends to become the expected doublet. With 3-O-glucosides and galactosides and 5-O-glucosides, the rotation of the glycosyl moiety is restricted and in the spectra of these flavonoids the C-1″ proton appears as a reasonably sharp doublet.

Flavonoid rhamnosides occur naturally as α-L-rhamnosides in which the rhamnosyl C-1″ proton has an equatorial-equatorial coupling with the C-2″ proton ($J = ca$ 2 c.p.s.). In both 3- and 7-O-rhamnosides the C-1″ proton signal occurs in the range 5·0–5·3 ppm (Table XI). In the NMR spectra of flavonoid rhamnosides, the rhamnose C-CH_3 proton signal (at 0·81–1·2 ppm) is also a distinguishing feature. The rhamnose methyl group in quercetin 3-O-rhamnoside gives a sharp doublet at 0·85 ppm ($J = 6·5$ c.p.s.) while in quercetin 3-glucoside-7-rhamnoside the rhamnose methyl group appears as a complex signal at 1·2 ppm. These differences observed in the coupling patterns probably reflect the freer rotation possible for a rhamnosyl moiety at C-7.

b. C-1″, C-1‴ and rhamnose methyl protons (C-6‴) in flavonoid rhamnosylglucosides. All the known flavonoid rhamnoglucosides contain either the rutinosyl or neohesperidosyl moieties (Fig. 16). These two substituents, which differ only in the point of attachment of the rhamnose to the glucose, can be distinguished in flavonoid disaccharides by the NMR analysis of their TMS ethers (Rösler *et al.* 1965). In the trimethylsilyl ethers of 7- and 3-O-rutinosides, rhamnose is characterized by a C-1‴ proton signal near 4·2–4·4 ppm ($J = 2$ c.p.s.) and a broad peak for the C-6‴ protons (methyl group) at 0·7–1·0 ppm, whereas in flavonoid 7- and 3-O-neohesperidosides, the rhamnose C-1‴ proton absorbs at 4·9–5·0 ppm ($J = 2$ c.p.s.) and the methyl group appears as a doublet at 1·1–1·3 ($J = 6$ c.p.s.).

5. *Methoxyl and Acetoxyl Protons*

In flavonoids, methoxyl proton signals appear in the region 3·7–4·1 ppm, while acetoxyl proton signals occur in the range 1·65–2·5 ppm. Signals for the protons on the flavonoid nucleus *ortho* and *para* to acetoxyl groups are shifted downfield by about 0·2–0·4 ppm relative to their positions in the spectrum of the flavonoid TMS ether; in contrast, *meta* proton signals are shifted downfield about 0·1 ppm.

A number of investigators (Massicot *et al.* 1963; Hillis and Horn, 1965; Eade *et al.* 1965; Horowitz and Gentili, 1966) have observed that the aromatic and aliphatic acetyl signals of *C*-glycosylflavonoid acetates fall in characteristic regions of the NMR spectra: the aromatic acetate signals appear between 2·30 and 2·50 ppm while the aliphatic acetate signals occur between 1·65 and 2·10 ppm. The signals of the 4′- and 7-acetyl groups in simple flavone acetates usually appear in the range of 2·30–2·35 ppm, while the signal for a 5-acetyl group occurs around 2·45 ppm.

In 8-*C*-glycosylflavones the 2″-O-acetyl signal consistently occurs in the

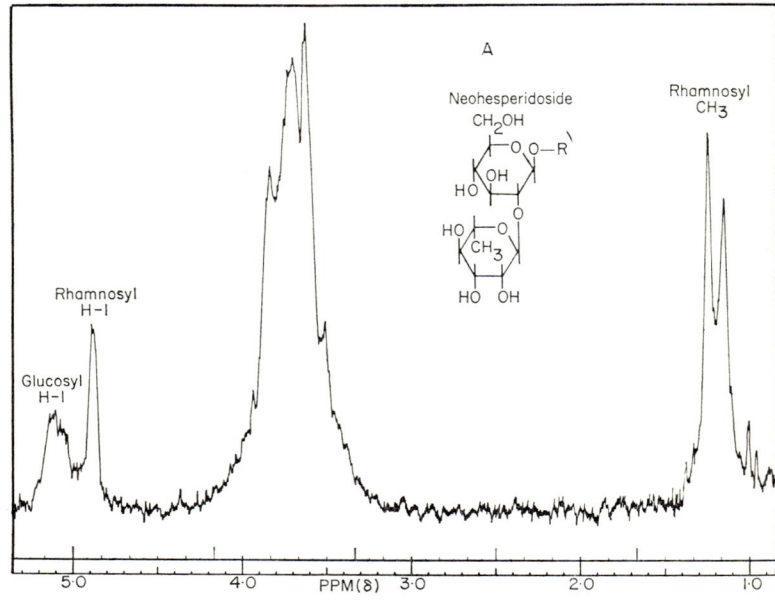

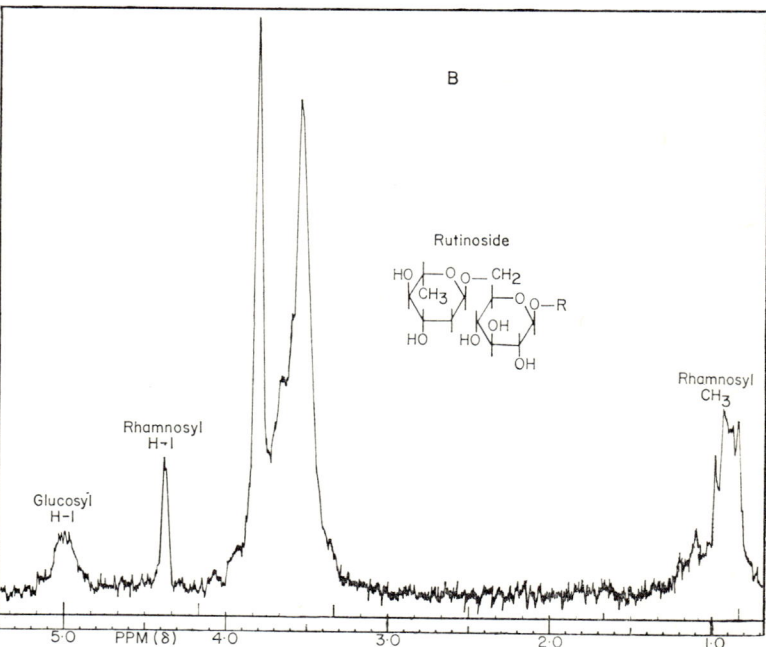

FIG. 16. NMR spectra of the sugar moiety in trimethylsilylated apigenin 7-O-neohesperido-side (A); and the sugar moiety in trimethylsilylated diosmetin 7-O-rutinoside (diosmin) (B).

range 1·70–1·75 ppm while the 6″-O-acetyl signal usually appears in the range 1·90–1·95 ppm; other acetyl signals associated with the C-glucosyl moiety come in the range 2·0–2·10 ppm. In 6-C-glycosylflavones, the signal for the 2″-O-acetyl group comes near 1·80 ppm and a band for the 6″-O-acetyl group appears near 2·0 ppm.

6. TMS Ether Protons

Most of the trimethylsilyl proton signals of flavonoid TMS ethers occur between 0·1 and 0·5 ppm. However the signals for a few trimethylsiloxy groups appear in the ranges 0 to 0·1 ppm upfield from tetramethylsilane (for example, the one at C-3 in dihydroflavonols and one in 8-C-glycosylflavones, probably at the 2″-position).

The number of hydroxyl groups present in a flavonoid can be determined by integrating the signals for the trimethyl-silyl protons.

APPENDIX

REAGENTS AND PROCEDURES FOR THE UV ANALYSIS OF FLAVONOIDS

Preparation of Reagent Stock Solutions and Solids

Sodium methoxide (NaOMe)   Freshly cut metallic sodium (2·5 g) was added cautiously in small portions to dry spectroscopic methanol (100 ml).

Aluminium chloride (AlCl₃)   Five grams of fresh anhydrous reagent grade $AlCl_3$ were added cautiously to spectroscopic methanol (100 ml).

Hydrochloric acid (HCl)   Concentrated reagent grade HCl (50 ml) was mixed with distilled water (100 ml).

Sodium acetate (NaOAc)   Anhydrous coarsely powdered reagent grade NaOAc was used.

Boric acid (H₃BO₃)   For Procedure I: Anhydrous powdered reagent grade H_3BO_3 was used.
For Procedure II: Spectroscopic methanol (100 ml) was saturated with anhydrous reagent grade H_3BO_3.

Steps in the UV Spectral Analysis

(a) A stock solution of the flavonoid was prepared by dissolving a small amount of the compound (less than half a mg) in about 10 ml of spectroscopic methanol. The concentration should be adjusted so that the absorptivity (optical density) of the major absorption peak between 250 and 400 nm lies in the region of 0·6 to 0·8 in a 1 cm cell.

(b) The methanol spectrum was measured using 2–3 ml of the stock solution of the flavonoid.

(c) *The NaOMe spectrum* was measured immediately after the addition of three drops of the NaOMe stock solution to the solution used for step *b*. After 5 min, the spectrum was rerun to check for flavonoid decomposition. The solution was then discarded.

(d) *The AlCl₃ spectrum* was measured immediately after the addition of six drops of the AlCl₃ stock solution to 2–3 ml of fresh stock solution of the flavonoid in methanol.

(e) *The AlCl₃/HCl spectrum* was recorded immediately after the addition of three drops of the stock HCl solution to the cuvette containing the AlCl₃ (from step *d*). The solution was then discarded.

(f) *The NaOAc spectrum* of the flavonoid was determined as follows. Coarsely powdered, anhydrous reagent grade NaOAc was added with shaking to 2–3 ml of fresh stock solution of flavonoid in a cuvette until an excess of NaOAc was present. In order to be certain that the solution was saturated, the cuvette was stoppered and set aside for about 2–5 min before measuring the spectrum. A second spectrum was run after 5–10 min to check for decomposition.

(g) *The NaOAc/H₃BO₃ spectrum* was determined as follows. Two methods for obtaining the NaOAc/H₃BO₃ spectra were used depending on whether or not decomposition of the flavonoid was observed during the recording of the NaOMe spectrum. If no decomposition was observed when the NaOMe spectrum was rerun after 5 minutes, procedure I was employed. When decomposition of the flavonoid in the presence of NaOMe did occur procedure II was used for the NaOAc/H₃BO₃ spectrum.

Procedure I: Sufficient powdered anhydrous reagent grade H₃BO₃ to give a saturated solution was added with shaking to the cuvette (from step *f*) which contained the NaOAc. The solution was discarded after the spectrum was recorded.

Procedure II: Five drops of the H₃BO₃ stock solution were added to 2–3 ml of fresh stock solution of the flavonoid. The solution was then quickly saturated with coarsely powdered reagent grade NaOAc and the NaOAc/H₃BO₃ spectrum was recorded.

The above procedures produced a set of six spectra for each of the 175 compounds examined; all the spectra are presented elsewhere (Mabry *et al.* 1969) in the exact form illustrated in Fig. 1.

REFERENCES

Batterham, T. S., and Highet, R. J. (1964). *Aust. J. Chem.* **17**, 428.

Dyke, S. F., Ollis, W. D., Sainsbury, M., and Schwarz, J. S. P. (1964). *Tetrahedron*, **20**, 1331.

Eade, R. A., Hillis, W. E., Horn, D. H. S., and Simes, J. J. H. (1965). *Aust. J. Chem.* **18**, 715.

Geissman, T. A., Jorgensen, E. L., and Harborne, J. B. (1953). *Chem. Ind. (London)*, 1389.

Grouiller, A. (1966). *Bull. Soc. Chim. France*, 2405.
Grouiller, A., and Pacheco, H. (1967). *Bull. Soc. Chim. France*, 1938.
Harborne, J. B. (1954). *Chem. Ind. (London)*, 1142.
Hillis, W. E., and Horn, D. H. S. (1965). *Aust. J. Chem.* **18**, 531.
Hörhammer, L., Hansel, R., and Strasser, R. (1952). *Arch. Pharm.* **285**, 438.
Hörhammer, L., Wagner, H., Rosprim, L., Mabry, T. J., and Rösler, H. (1965). *Tetrahedron Lett.* 1707.
Horowitz, R. M. and Gentili, B. (1966). *Chem. Ind.* **625**.
Jurd, L. (1962). *In* "The Chemistry of Flavonoid Compounds" (T. A. Geissman, Ed.), pp. 107–155, Pergamon Press, Oxford.
Jurd, L., and Geissman, T. A. (1956). *J. org. Chem.* **21**, 395.
Jurd, L., and Horowitz, R. M. (1961). *J. org. Chem.* **26**, 2561.
Mabry, T. J., Kagan, J., and Rösler, H. (1965a). *Phytochemistry*, **177**, 487.
Mabry, T. J., Kagan, J., and Rösler, H. (1965b). The University of Texas Publication No. 6418.
Mabry, T. J., Markham, K. R., and Thomas, M. B. (1969). "The Systematic Identification of Flavonoids", Springer-Verlag New York Inc., New York, U.S.A.
Markham, K. R., and Mabry, T. J. (1968). *Phytochemistry*, **7**, 1197.
Markham, K. R., Swift III, W. T., and Mabry, T. J. (1967). *J. org. Chem.* **53**, 462.
Massicot, J., and Marthe, J.-P. (1962). *Bull. Soc. Chim. France*, 1962.
Massicot, J., Marthe, J.-P., and Neitz, S. (1963). *Bull. Soc. Chim. France*, 2712.
Narasimhachari, N., and Seshadri, T. R. (1948). *Proc. Indian Acad. Sci.* **27A**, 223.
Narasimhachari, N., and Seshadri, T. R. (1951). *Proc. Indian Acad. Sci.* **30A**, 271.
Rösler, H., Mabry, T. J., Cranmer, M. F., and Kagan, J. (1965). *J. org. Chem.* **30**, 4346.
Rösler, H., Mabry, T. J., and Kagan, J. (1965). *Chem. Ber.*, **98**, 2193.
Seikel, M. K., and Mabry, T. J. (1965). *Tetrahedron Lett.* 1105.
Swain, T. (1954). *Chem. Ind. (London)*, 1480.
Waiss, A. C., Lundin, R. E., and Stern, D. J. (1967). *Tetrahedron Lett.* 513.

CHAPTER 2

The Biosynthesis of Cyanogenic Glycosides and Other Simple Nitrogen Compounds

ERIC E. CONN

Department of Biochemistry and Biophysics, University of California at Davis

and

G. W. BUTLER

Plant Chemistry Division, Division of Scientific and Industrial Research, Palmerston North, New Zealand

I. Introduction	. . . .	47
II. Early Studies on Biosynthesis	. . . .	50
III. Recent Studies on Biosynthesis	. . . . . .	57
IV. Biosynthesis of Mustard Oil Glucosides	. . . . . .	63
V. Metabolism of Cyanide	. . . .	66
VI. Conclusion	. . . .	70
References	. . . .	72

I. Introduction

The ability of plants to produce HCN from a parent substance was first reported in 1803 by a German pharmacist (Schrader, 1803). This property, known as cyanogenesis, is exhibited by at least 750 species representing approximately 60 families and 250 genera (Hegnauer, 1963). Although the parent substance in the majority of these cyanophoric plants has not been identified, it is probably one of the 11 cyanogenic glycosides described in Table I. It will be noted that the aglycone of the glycoside in every case is an α-hydroxynitrile (cyanohydrin). The older literature on the cyanogenic glycosides has been reviewed by Robinson (1930) and Dilleman (1958).

During the past decade several laboratories have examined the biosynthesis of this group of compounds. One of the present authors (G.W.B.) and his associates have been primarily concerned with linamarin (α-hydroxyisobutyronitrile-β-glucopyranoside) (I) and its homologue, lotaustralin. These are the two glycosides with aliphatic aglycones which are widely distributed in cyanophoric plants, both compounds usually being present in any single species

TABLE I

Cyanogenic glycosides of known structure

Glycoside	Sugar	Aglycone	Occurrence
Linamarin	D-Glucose	α-Hydroxyisobutyronitrile (acetone cyanohydrin)	*Linum usitatissimum, Phaseolus lunatus, Trifolium repens, Lotus* sp.
Lotaustralin	D-Glucose	α-Hydroxy-α methyl butyronitrile (methyl ethyl ketone cyanohydrin)	(See linamarin)
Acacipetalin	D-Glucose	β-Dimethyl-α-hydroxyacrylonitrile	*Acacia* sp. (South African)
Prunasin	D-Glucose	D-Mandelonitrile	*Prunus* sp., many Rosaceae, *Eucalyptus* sp.
Sambunigrin	D-Glucose	L-Mandelonitrile	*Sambucus nigra, Acacia* sp. (Australian)
Prulaurasin	D-Glucose	DL-Mandelonitrile[a]	
Amygdalin	Gentiobiose	D-Mandelonitrile	*Prunus* sp.
Vicianin	Vicianose	D-Mandelonitrile	*Vicia angustifolia* and other *Vicia*
Dhurrin	D-Glucose	L-*p*-Hydroxymandelonitrile	*Sorghum* sp.
Taxiphyllin	D-Glucose	D-*p*-Hydroxymandelonitrile	*Taxus* sp.
Zierin	D-Glucose	*m*-Hydroxymandelonitrile	*Zieria laevigata*
Gynocardin[b]	D-Glucose	Gynocardinonitrile	*Gynocardia odorata, Pangium edule*

See Dilleman (1958) for references.

[a] Prulaurasin is undoubtedly an artifact of isolation (Plouvier, 1935).
[b] Gynocardin structure proposed by Coburn and Long (1966).

(Butler, 1965). The biosynthesis of linamarin has been studied in germinating seedlings of the linen flax *Linum usitatissimum* L. in which a rapid synthesis occurs.

Dhurrin (*p*-hydroxy-L-mandelonitrile-β-glucopyranoside) (II) the cyanogen of *Sorghum vulgare* Pers. was of particular interest to the other author (E.E.C) because compounds with its phenylethane carbon skeleton are relatively rare

Linamarin   Dhurrin

(I)   (II)

in nature (Geissman and Hinreiner, 1952). This glucoside is absent in the dry sorghum seed but is synthesized rapidly and reaches a concentration of 3–5% (dry weight) in three-day-old dark-grown seedlings (Akazawa *et al.* 1960). The sorghum seedling, therefore, also readily lends itself to biosynthetic studies.

Since their discovery, considerable effort has been expended studying the catabolism of cyanogenic glycosides. It is informative to summarize this work because occasionally a slight modification in a degradation scheme for a compound may account for its biosynthesis. Figure 1 describes, as an example, the events that occur when sorghum tissue containing dhurrin is disrupted. Initially, the dhurrin is hydrolysed by a β-glucosidase present in the plant and *p*-hydroxy-L-mandelonitrile and D-glucose are formed. The glucosidases that act on the cyanogens in sorghum (Mao and Anderson, 1967) and flax (Butler *et al.* 1965) have been partially purified and studied. These plants contain two β-glucosidases, one that catalyses the hydrolysis of the cyanogens and another that is active towards salicin or other substrates. Two β-glucosidases in sweet almonds have also been crystallized and their properties reported (Helferich and Kleinschmidt, 1968).

While the cyanohydrin formed by the action of the glucosidase can dissociate non-enzymically to yield HCN and an aldehyde or ketone (Fig. 1), the reaction is also catalysed by an enzyme present in the sorghum seedling. This type of enzymatic activity in almonds was first described by Rosenthaler (1908) who was interested in the reverse process, namely the synthesis of benzaldehyde cyanohydrin from HCN and benzaldehyde. Rosenthaler's work was one of the earliest examples of asymmetric synthesis catalysed by an enzyme and it stimulated much additional work (Albers, 1941). The enzyme in sorghum, hydroxynitrile lyase, which catalyses the cyanohydrin equilibrium has been obtained as a homogeneous protein (Seely *et al.* 1966) and its properties have been described. A similar enzyme has been purified from the seed of bitter almonds (Becker *et al.* 1963). There is one surprising difference between the enzymes isolated from these two plant sources. The almond enzyme contains a flavin prosthetic group, FAD, which is required for enzymatic activity while the highly purified sorghum enzyme is devoid of flavin. Since the reaction catalysed does not involve oxidation or reduction, the role of a flavin is not well understood.

The possibility of the cyanogenic glucosides being formed by a reversal of the degradative sequence in Fig. 1 was examined at an early date in our studies (Koukol *et al.* 1962; Butler and Conn, 1964a). Specifically, HCN-^{14}C was

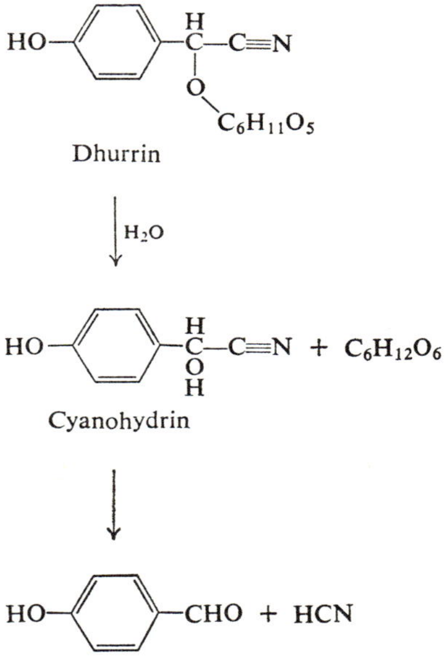

FIG. 1. The enzymatic destruction of dhurrin by sorghum enzymes.

administered as a gas to sorghum and flax seedlings on the assumption that radioactivity would be incorporated into the cyanogenic glucosides if synthesis were due to a reversal of the degradation sequence. Although the HCN-^{14}C was extensively metabolized (Blumenthal-Goldschmidt *et al.* 1963) in both flax and sorghum seedlings, there appeared to be no significant labelling of the cyanogenic glucosides. For this reason the possibility of the aglycones being produced from cyanide and the corresponding aldehyde or ketone was discounted in our early studies.

II. EARLY STUDIES ON BIOSYNTHESIS

Ten years ago to the month that this review was presented at the Anniversary Meeting of the Phytochemical Society, two papers read at the Federation Meetings in the United States reported independent studies on the origin of the aglycone of dhurrin in etiolated sorghum seedlings (Gander, 1958; Conn and Akazawa, 1958). In his study, Gander had concerned himself with the origin of the nitrile carbon of the glucoside, while Conn and Akazawa, because of their interest in aromatic metabolism, investigated the precursors for the p-hydroxybenzaldehyde moiety. Table II shows the data on the origin of the aldehyde moiety presented at those meetings. L-Tyrosine-U-^{14}C and its precursor shikimic acid-U-^{14}C were effectively converted into the aldehyde moiety of dhurrin. Glucose-U-^{14}C was about one-tenth as effective, and acetate-2-^{14}C contributed only insignificantly to labelling in the aldehyde moiety. In these experiments, DL-phenylalanine-3-^{14}C did not significantly label the aldehyde moiety, thereby indicating that, as in most plants (Neish, 1964), tyrosine is not readily synthesized from phenylalanine in sorghum.

TABLE II

Origin of the Aldehyde moiety of dhurrin

Compound administered	Amount fed (μM/g)	Uptake by plant %	Dilution factor	Incorporation %
Glucose-U-^{14}C	10·8	85	520	0·43
Shikimic-U-^{14}C	12·4	62	46	4·23
L-Tyrosine-U-^{14}C	10·1	89	55	3·65
DL-Phenylalanine-3-^{14}C	12·0	92	3600	0·05
Acetate-2-^{14}C	15·3	93	1300	0·11

Data are from Koukol et al. (1962).

In his paper, Gander reported that DL-tyrosine-2-^{14}C gave rise to labelled cyanide obtained from dhurrin, whereas carboxyl-labelled and 3-labelled tyrosine did not.

Following the discovery that the two laboratories were studying the same problem, there was a profitable exchange of information and a number of other experiments were designed to identify intermediates in the conversion of tyrosine to dhurrin. It seemed reasonable to expect that such intermediates could be detected and identified because the amount of radioactivity incorporated was large. As an example, when 10 micromoles of radioactive tyrosine are administered to a gram of sorghum seedlings, it is routine to observe that 5 to 15% of the carbon-14 in the tyrosine is converted into dhurrin in 24 to 48 h. It is interesting to note that, although the experimental conditions under which the seedlings were grown and fed tyrosine-^{14}C were varied extensively, no significantly labelled compounds other than tyrosine and dhurrin could be observed in the alcohol soluble fraction (Gander, 1960; Gander, 1962).

When no compounds which might possibly be intermediates were detected by this approach, the next step was to synthesize possible intermediates containing carbon-14, administer these to intact sorghum seedlings through their roots and observe whether or not they were incorporated. This approach then ruled out the following compounds as precursors of dhurrin under the conditions employed: tyramine-U-^{14}C; tyramine-8-^{14}C; p-hydroxyphenylacetic acid-U-^{14}C; $trans$-cinnamic acid-3 and ring-^{14}C; and p-hydroxycinnamic (p-coumaric) acid-3-^{14}C (Koukol $et al.$ 1962; Gander, 1962). On the other hand p-hydroxyphenylpyruvic acid-3-^{14}C, p-hydroxyphenyl-DL-lactic acid-3-^{14}C and p-hydroxybenzaldehyde-7-^{14}C were reported to be effectively converted to dhurrin (Koukol $et al.$ 1962). The ability of the first two compounds to serve as precursors was attributed to the ease with which they could be converted to tyrosine in the intact seedling. In view of later studies on the origin of the nitrile nitrogen atom of dhurrin, this appears to have been an accurate interpretation.

The ability of p-hydroxybenzaldehyde-7-^{14}C to serve as a precursor of dhurrin was difficult to interpret in view of the fact the HCN-^{14}C fed to the same species did not label the cyanogenic glucoside. Subsequent studies (Libby, 1964) showed that p-hydroxybenzaldehyde-7-^{14}C administered to sorghum seedlings was converted to a glucoside, presumably p-glucosyloxybenzaldehyde, which had chromatographic properties quite similar to dhurrin but which, in contrast to the cyanogen, was confined to the roots of the seedlings. In a simple experiment, Libby showed that the labelled aldehyde did not serve as a precursor when the tops of seedlings fed p-hydroxybenzaldehyde-7-^{14}C were separated prior to isolation of dhurrin. Since nearly all of the dhurrin in young seedlings is found in the tops, this readily demonstrated that the aldehyde did not significantly label the cyanogen.

In these early studies, a β-hydroxy-α-aminophenylpropionic acid was considered as a possible intermediate because the initial step in modification of a precursor amino acid could be oxidation at its β-position. In particular, Gander (1960) had considered β-oxidation as a first step in the formation of dhurrin from tyrosine. While his results indicated that p-hydroxyphenylserine did increase the conversion of tyrosine to dhurrin in whole seedlings, phenylserine-U-^{14}C and glycine-2-^{14}C did not significantly label dhurrin in later feeding experiments (Gander, 1962). Koukol $et al.$ (1962) also examined the possibility of glycine-2-^{14}C condensing with p-hydroxybenzaldehyde $in vivo$ as it is known to occur in animal tissues (Bruns and Fiedler, 1958) to form p-hydroxyphenylserine-2-^{14}C. These experiments were also negative. Finally when a mixture of the four isomers of radioactive p-hydroxyphenylserine was fed to sorghum plants, it did not significantly label dhurrin nor did the mixtures of the $erythro$ and $threo$ isomers when fed separately (Uribe, 1965).

Still another approach involving competition experiments in which radioactive tyrosine was administered together with inactive compounds that might be intermediates, has yielded no encouraging information regarding the biosynthetic sequence for dhurrin (Uribe, 1965).

Compounds that similarly might be considered as intermediates in the conversion of valine to linamarin by flax were examined by Butler and Conn (1964a). Known precursors of valine were included in this study but there was no light shed on the nature of the intermediates involved in linamarin biosynthesis using ^{14}C-labelled compounds. In these studies it was possible to greatly increase the conversion of labelled valine to linamarin by using only the tops of the flax seedlings rather than the entire seedling. With only the tops, the period for uptake of the labelled precursor could be shortened to about an hour and, after a total period of 7 h, 35–50% of the carbon-14 administered as valine was converted to linamarin.

The ability to conduct short-term experiments with seedling tops encouraged the authors to perform competition experiments with flax seedlings also. Again the possibility of an early step being the oxidation of the precursor amino acid at the β-position was considered. Competition experiments with DL-β-hydroxyvaline and α-hydroxyisobutyric acid did show that the conversion of valine-U-^{14}C to linamarin was inhibited. However, when α-hydroxyisobutyric acid-1-^{14}C and DL-β-hydroxyvaline-^{3}H were administered directly to seedling tops, only slight incorporation of radioactivity into linamarin occurred. Tschiersch (1966a, b) has recently confirmed our results with β-hydroxyvaline.

By 1964, the ability of other amino acids to serve as precursors of several other cyanogenic glucosides had been demonstrated. Thus, Butler and Butler (1960) reported that valine and isoleucine were effective precursors of linamarin and lotaustralin in white clover. Phenylalanine was shown to be a precursor of prunasin in cherry laurel leaves in 1961 (Mentzer and Favre-Bonvin, 1961) and later in peach seedlings (Ben-Yehoshua and Conn, 1964). The same structural relationship between the amino acid and the cyanogenic glucoside was indicated by these studies. That is, the conversion of the amino acid to the cyanogenic glucoside must involve a reaction sequence in which the carboxyl carbon of the amino acid is lost, the alpha carbon is oxidized to the level of a nitrile and the beta carbon acquires a hydroxyl group that is subsequently glucosylated to yield the cyanogenic glucoside (Fig. 2). While these reactions would account for the conversion of the amino acid to the aglycone, there was no indication in which order these modifications might occur.

In view of the recent developments in this problem it is important to note that the French workers (Mentzer and Favre-Bonvin, 1961; Mentzer et al. 1963) suggested that the oxime of β-hydroxyphenylpyruvic acid (a ketoxime) might be an intermediate in the formation of prunasin from phenylalanine in the cherry laurel. The oxime was proposed because of the facile conversion (Fig. 3) of this class of compounds to nitriles (Ahmad and Spenser, 1961). Although Ahmad and Spencer, who have studied this reaction extensively, wrote a concerted mechanism involving simultaneous decarboxylation and dehydration, the conversion is written in Fig. 3 in two separate steps with an aldoxime as an intermediate. Reasons for doing so will become clear as later studies in this problem are reviewed.

3

When Mentzer and Favre-Bonvin proposed that a ketoxime might be an intermediate, they suggested that this compound would be formed from the keto-acid analogue of an amino acid and NH_2OH. For this reason, but also in

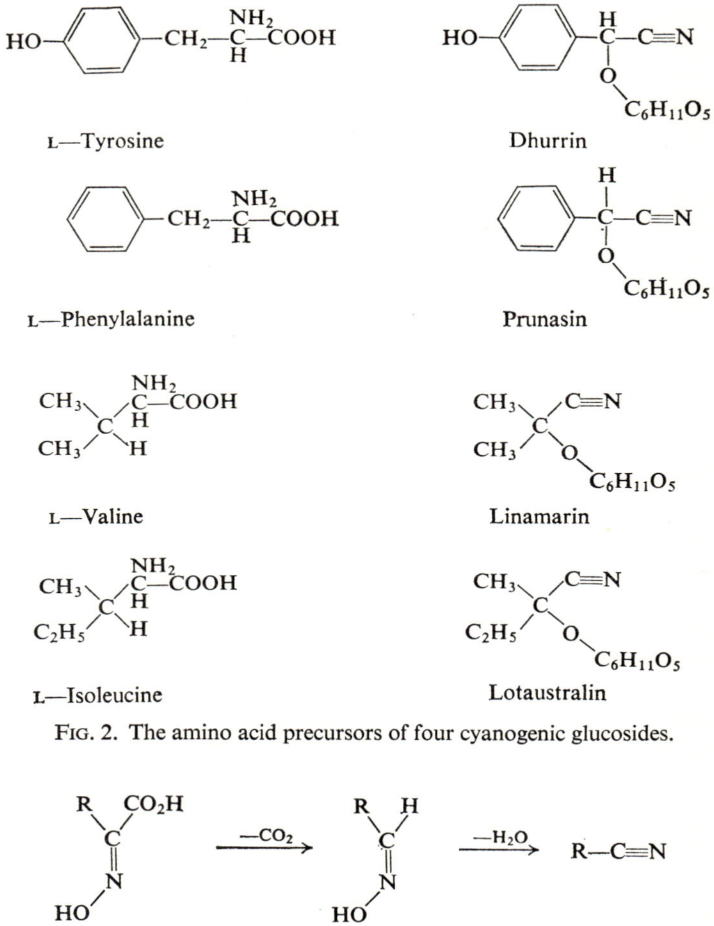

FIG. 2. The amino acid precursors of four cyanogenic glucosides.

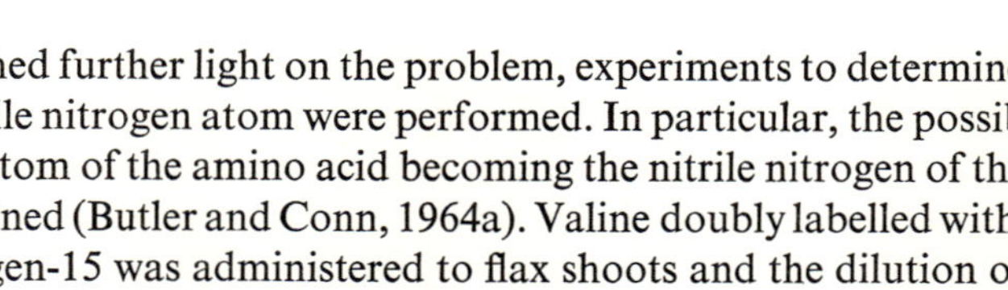

FIG. 3. The conversion of α-keto acid oximes to nitriles.

order to shed further light on the problem, experiments to determine the origin of the nitrile nitrogen atom were performed. In particular, the possibility of the nitrogen atom of the amino acid becoming the nitrile nitrogen of the cyanogen was examined (Butler and Conn, 1964a). Valine doubly labelled with carbon-14 and nitrogen-15 was administered to flax shoots and the dilution of these two isotopes was measured as conversion to linamarin occurred. In two experiments (Table III) the carbon isotope was diluted 3·9 and 7·6 fold, whereas in the same experiments the nitrogen isotope was diluted 6·3 and 17·2. Inasmuch as the amino acid would be expected to undergo transamination during the course of the experiment and therefore loss of nitrogen-15, these data were interpreted as showing that the nitrogen atom of the cyanogenic glucoside is directly derived from the precursor amino acid. Data for a similar experiment in the case of dhurrin biosynthesis in sorghum seedlings have been obtained (Uribe and Conn, 1966) (Table IV). In these experiments the dilution of the nitrogen-15 was only about 20% greater than the dilution of the carbon isotope. These data clearly indicated that the nitrogen is derived directly from the

TABLE III

Dilution of ^{14}C and ^{15}N on incorporation of L-valine-^{14}C-^{15}N into linamarin-^{14}C-^{15}N

Expt. No.	^{15}N Concentration (apxs)		^{14}C Concentration (MμC/μatom C)	
	Val	Linamarin	Val-α-C	Linamarin-CN
1	45·4	7·17	37·8	9·6
2	45·4	2·64	39·8	5·2

Expt. No.	Dilution of		Ratio ^{14}C:^{15}N	
	^{14}C	^{15}N	Calc.	Found
1	3·9	6·3	1·00	1·60
2	7·6	17·2	1·00	2·23

Data from Butler and Conn (1964a).

precursor amino acid and made it mandatory that all intermediates between the amino acid and the aglycone be nitrogenous in nature. Bleichert *et al.* (1966) have conducted a similar experiment in the biosynthesis of taxiphyllin in *Taxus* from doubly labelled ^{14}C-^{15}N-tyrosine. In this plant the nitrogen of the amino acid was also retained when tyrosine was converted to the cyanogen.

Another experiment that placed limitations on the nature of the intermediates between the precursor amino acid and cyanogenic glucoside was performed by Koukol *et al.* (1962). In order to determine whether the bond between the C-2 and C-3 atoms of tyrosine is broken as tyrosine is converted to dhurrin, sorghum seedlings were fed tyrosine labelled in the 2- and the 3-positions in known ratios. The ratio of the specific activities of *p*-hydroxy-benzaldehyde and HCN obtained from the biosynthesized dhurrin was subsequently determined (Table V). The ratios of the two labelled positions were

TABLE IV

Dilution of ^{14}C and ^{15}N on incorporation of L-tyrosine-^{14}C-^{15}N into dhurrin-^{14}C-^{15}N

Expt. No.	^{15}N concentration (apxs)		^{14}C concentration (cpm/μatom C)	
	Tyr	Dhurrin	Tyr-α-C	Dhurrin-CN
1	50·03	1·17	20·800	610
2	50·03	1·18	42·800	1220

Expt. No.	Dilution of		Ratio ^{14}C:^{15}N	
	^{14}C	^{15}N	Calc.	Found
1	34	43	1·00	1·26
2	35	42	2·00	2·40

Data from Uribe and Conn (1966).

TABLE V

Synthesis of dhurrin from doubly-labelled tyrosine

Expt. No.	Compounds administered[a]		Compounds isolated	
	L-Tyrosine-3-C^{14} / L-Tyrosine-2-C^{14}		p-Hydroxybenzaldehyde / HCN	
		Ratio		Ratio
3A	0·0172 μC/μmole / 0·0344 μC/μmole	1 / 2	0·0022 μC/μmole / 0·0041 μC/μmole	1 / 1·9
3B	0·0513 μC/μmole / 0·0171 μC/μmole	3 / 1	0·0058 μC/μmole / 0·0020 μC/μmole	2·9 / 1
4	0·0344 μC/μmole / 0·0344 μC/μmole	1 / 1	0·0034 μC/μmole / 0·0022 μC/μmole	1·54 / 1

Data from Koukol et al. (1962).
[a] Administered as DL-Tyrosine.

effectively maintained in three separate experiments (1 : 2; 3 : 1; 1 : 1). Therefore, it was concluded that the carbon–carbon bond between the α and β carbon atoms is not cleaved as the cyanogenic glucoside is formed. These results appeared to rule out the possibility of the alpha carbon of tyrosine being in some manner converted to HCN, the ring and β-carbon atoms of the amino acid giving rise to p-hydroxybenzaldehyde, and these two moieties condensing to yield a cyanohydrin that could be glucosylated. In the biosynthesis of taxiphyllin from tyrosine in *Taxus*, Bleichert *et al.* (1966) similarly have shown that the bond between C-2 and C-3 remains intact.

III. RECENT STUDIES ON BIOSYNTHESIS

The experiments described in the preceding section indicated that only nitrogenous intermediates were involved in the biosynthesis of the cyanogenic glucosides. While this appeared to rule out the formation of an oxime intermediate (Mentzer *et al.* 1963), the conversion of 3-indole acetaldoxime to 3-indole acetonitrile in several fungi and banana reported by Mahadevan

TABLE VI

Conversion of oximes to linamarin

			Linamarin	
Compound administered	Amount (μmoles)	Sp. act. (μC/mmole)	Dilution factor	$\%C^{14}$ converted
L-Valine-U-^{14}C[a]	1·2	730	43	23
α-Keto-isovaleric acid Oxime-U-^{14}C[a]	1·0	920	120	9
L-Valine-U-^{14}C[a]	3·3	240	18	25
Isobutyraldoxime-U-^{14}C	3·4	390	21	21
Isobutyraldehyde-U-^{14}C	1·3	1510	1510	0·7

Data from Tapper *et al.* (1967).
[a] Specific activity corrected for loss of carboxyl carbon atom.

(1963) required that oximes be examined as intermediates. Further, an enzyme system that converts 3-indole acetaldoxime to 3-indole acetonitrile has been described (Kumar and Mahadevan, 1963). Therefore the oximes of α-ketoisovaleric acid (2-oximinovaline) and isobutyraldehyde containing carbon-14 were synthesized and administered to flax plants (Tapper *et al.* 1967). The data obtained (Table VI) showed that α-ketoisovaleric acid oxime-U-^{14}C, while only about half as effective as the amino acid itself, is nevertheless converted into linamarin. Isobutyraldoxime-U-^{14}C was as effective a precursor as valine itself. The corresponding aldehyde, derivable by hydrolysis of the aldoxime, did not serve as a precursor.

Since α-ketoisovaleric acid oxime-U-^{14}C might be initially converted to valine in the plant through hydrolysis to the keto acid and subsequent transamination, it was important to determine whether or not the nitrogen atom of the ketoxime was incorporated into linamarin. Accordingly α-ketoisovaleric acid oxime doubly labelled with ^{14}C and ^{15}N has been synthesized and administered to flax plants (B. A. Tapper, unpublished observations). Analysis of linamarin for ^{14}C and ^{15}N showed that both labels were diluted approximately equally on conversion to the glucoside. Similar results were obtained with doubly labelled isobutyraldoxime.

The ability of oximes to act as effective precursors of linamarin in flax seedlings was considered sufficiently unique to require that oximes be tested as precursors in the biosynthesis of another cyanogenic glucoside. Phenylpyruvic acid oxime-2-^{14}C and phenylacetaldoxime-U-^{14}C were synthesized and administered to shoots and single leaves of cherry laurel (*Prunus laurocerasus* L.) (B. A. Tapper, unpublished observations; Hahlbrock *et al.* 1968). The data obtained (Table VII) confirmed that the oxime of the keto acid is

TABLE VII

Conversion of oximes to prunasin

Compound administered	Sp. act. (μC/mmole)	Sp. act. of products isolated (μC/mmole)		%^{14}C converted
		HCN	Benzaldehyde	
L-Phenylalanine-U-$^{14}C^a$	439	0·032	0·22	8·3
Phenylpyruvic acid Oxime-2-^{14}C	271	2·23	< 0·002	41
L-Phenylalanine-U-$^{14}C^a$	439	0·026	0·176	16
Phenylacetaldoxime-U-^{14}C	23	0·004	0·033	43

Data from Tapper (unpublished).
[a] Specific activity corrected for loss of carboxyl carbon atom.

efficiently converted to prunasin in this species. By comparison with L-phenylalanine-U-^{14}C, the ketoxime was five times more effectively converted, while phenylacetaldoxime-U-^{14}C was also extensively converted to prunasin.

These results have rapidly led to additional experiments designed to identify the intermediates between the aldoximes and the cyanogenic glucosides. Two possibilities obviously presented themselves here (Fig. 4) and are illustrated with isobutyraldoxime as an example. In one case, the branch carbon atom could be first oxidized and glucosylated before the oxime group is dehydrated to yield the nitrile. An alternative route is the conversion of the oxime to the nitrile by dehydration, then oxidation of the nitrile to the α-hydroxynitrile (cyanohydrin) and subsequent glucosylation. It is this second possibility that has been examined experimentally because of the comparative ease with which these compounds could be labelled with carbon-14 and the difficulty encountered in synthesizing the oxime of the α-hydroxyisobutyraldehyde indicated in Fig. 4.

When isobutyronitrile-1-^{14}C and α-hydroxyisobutyronitrile-1-^{14}C were administered to flax shoots (Table VIII), they were observed to be effective precursors of linamarin (B. A. Tapper, unpublished observations; Hahlbrock *et al.* 1968). In the second of two experiments in which α-hydroxyisobutyronitrile-1-^{14}C was fed, the radioactivity of the nitrile group of the cyanohydrin was incorporated into the nitrile carbon of linamarin without randomization of the isotope.

Similar data were obtained when phenylacetonitrile-1-^{14}C and α-hydroxyphenylacetonitrile-1-[14]C (mandelonitrile-1-^{14}C) were examined as possible precursors of prunasin in cherry laurel shoots and leaves (Table IX). Here, the

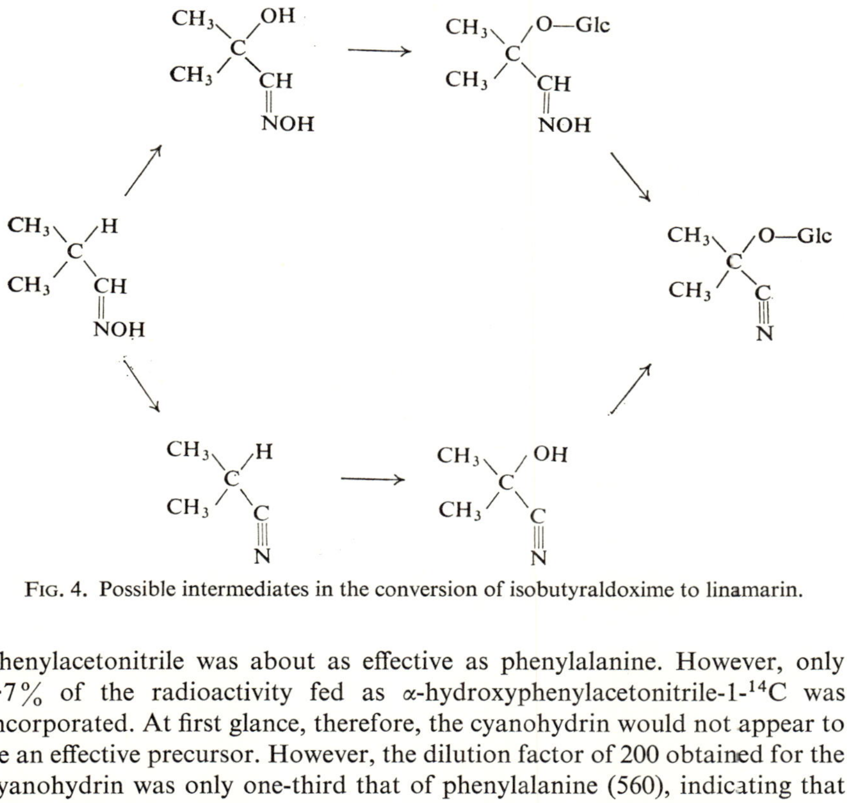

Fig. 4. Possible intermediates in the conversion of isobutyraldoxime to linamarin.

phenylacetonitrile was about as effective as phenylalanine. However, only 0·7% of the radioactivity fed as α-hydroxyphenylacetonitrile-1-^{14}C was incorporated. At first glance, therefore, the cyanohydrin would not appear to be an effective precursor. However, the dilution factor of 200 obtained for the cyanohydrin was only one-third that of phenylalanine (560), indicating that the cyanohydrin is a precursor. In this experiment, somewhat larger amounts of the mandelonitrile were administered because of the expected dissociation of the cyanohydrin to yield HCN. The poor incorporation observed here has been attributed to the toxic effect of cyanide in equilibrium with the high concentration of cyanohydrin fed. Indeed, the cherry laurel shoots became necrotic a few hours after immersion in the solution containing the mandelonitrile. Even though lower concentrations of mandelonitrile have been tried,

TABLE VIII

Conversion of nitriles to linamarin

Compound administered	Linamarin			
	Dilution factor	% ^{14}C converted	% ^{14}C as CN	
			Found	Calculated
L-Valine-U-^{14}C[a]	55	23		
Isobutyronitrile-1-^{14}C	10	11		
α-Hydroxyisobutyro-nitrile-1-^{14}C	17	28		
L-Valine-U-^{14}C[a]	7×10^4	28	26	25
α-Hydroxyisobutyro-nitrile-1-^{14}C	2·4	8	100	100

Data from Hahlbrock et al. (1968).
[a] Corrected for loss of carboxyl carbon atom.

TABLE IX

Conversion of nitriles to prunasin

Compound administered	Prunasin			
	Dilution factor	% ^{14}C converted	% ^{14}C as CN	
			Found	Calculated
DL-Phenylalanine-2-^{14}C	290	25	92	100
L-Phenylalanine-U-^{14}C[a]	620	27	16	12·5
Phenylacetonitrile-1-^{14}C	210	19	94	100
DL-Phenylalanine-2-^{14}C	560	12	87	100
α-Hydroxyphenylaceto-nitrile-1-^{14}C	200	0·7	95	100

Data from Hahlbrock et al. (1968).
[a] Corrected for loss of carboxyl carbon atom.

the plants in general are injured when immersed in solutions of this cyano-hydrin.

When some cyanohydrins were found to be effective precursors of the cyano-genic glucosides, it seemed desirable to re-examine the experiments in which HCN was eliminated as a precursor. Further study showed that the results obtained are dependent on the manner in which the cyanide is fed. In the early experiments in which incorporation into dhurrin was not observed (Blumenthal-Goldschmidt *et al.* 1963), cyanide was administered as a gas enclosed in a vessel containing sorghum seedlings. More recently, Tapper (unpublished) has administered HCN-^{14}C in solution with and without acetone and measured the incorporation of radioactivity into linamarin in flax seedlings. When cyanide and acetone are incubated together for a short period before the mixture is administered to the plants, radioactivity from HCN-^{14}C is incor-

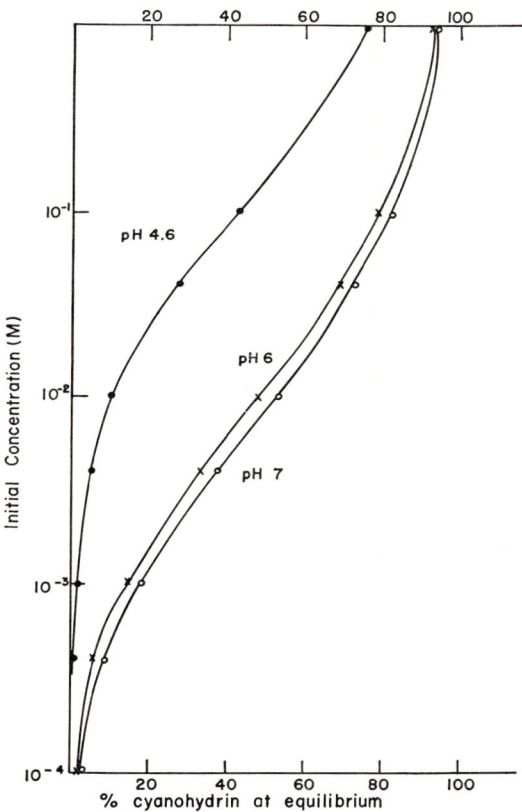

FIG. 5. The dissociation of acetone cyanohydrin as a function of initial concentration.

porated into linamarin. This period of time, of course, permits cyanide and acetone to associate and form the cyanohydrin. When the latter is present in the feeding solution, it then can serve as a precursor of linamarin.

A. Bauer (unpublished results) has measured the equilibrium constant for acetone cyanohydrin formation in aqueous solutions at three different pH values. From these constants, one can calculate the extent to which the cyanohydrin will dissociate as a function of its initial concentration in aqueous solution (Fig. 5). At pH 7, for example, about 55% of the cyanohydrin will

3*

remain associated at equilibrium when the initial concentration of the cyano-hydrin is 0·01 M. If higher concentrations are prepared, even more cyano-hydrin will remain at equilibrium, whereas at lower initial concentrations the cyanohydrin dissociates extensively. Whether equilibrium is obtained within the flax plant during the course of the experiment is not known, even though an enzyme that can catalyze this cyanohydrin equilibrium is known to be present.

To summarize, these new results indicate that, in the case of two cyanogenic glucosides, the biosynthetic pathway leads from amino acid to cyanogen through a series of nitrogenous intermediates including oximes and nitriles (Fig. 6). Experiments of this sort are open to familiar criticisms, however. For

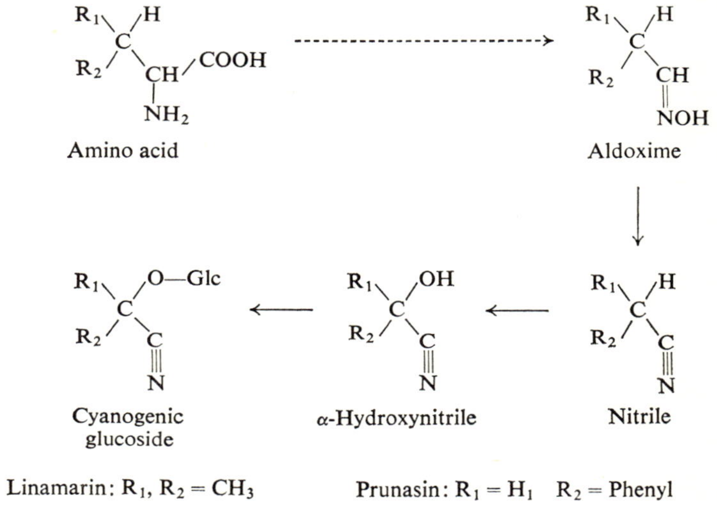

FIG. 6. A possible, partial biosynthetic pathway from amino acid to cyanogenic glucoside.

example, two compounds whose effectiveness as precursors is being compared may not be absorbed and transported to the site of biosynthesis at equal rates. Or, a natural compound may be toxic when administered in large amounts compared with the low concentration normally present in the plant. Further, a labelled compound may be incorporated and converted to the product under study even though it is not a natural intermediate. This is a possibility in the case of the cyanohydrins employed in the recent work, in view of the well-known ability of plants to glucosylate compounds containing hydroxyl groups (Pridham and Saltmarsh, 1963). To establish a biosynthetic pathway conclu-sively, the intermediates proposed should be shown to be produced and utilized by enzymes catalysing separate reactions in the pathway.

In the present study it has been possible by means of trapping experiments to obtain evidence for an oxime as an intermediate. In the biosynthesis of linamarin, Tapper (unpublished) has fed valine-U-^{14}C and non-radioactive isobutyraldoxime simultaneously to flax seedling tops and subsequently isolated isobutyraldoxime from the plant material. When this was done, the purified isobutyraldoxime was found to be radioactive. Thus, valine-U-^{14}C had been converted by the plants to isobutyraldoxime under these conditions.

Further support for oxime intermediates is found in experiments in which valine-U-^{14}C was fed to flax shoots together with an amino acid analogue, O-methylthreonine. Under these conditions, the conversion of valine to linamarin decreased and labelled isobutyraldoxime and an unidentified labelled glucoside were formed. The latter compound yields isobutyraldehyde under mild acid conditions and is believed to be the glucoside of isobutyraldoxime (Tapper, unpublished). Gander (1966) has also described an unidentified glucoside in *Sorghum vulgare* seedlings that may be a precursor of dhurrin in that species.

The oximes, nitriles and cyanohydrins reported in Tables VI to IX to be converted to cyanogenic glucosides were about equally effective as precursors. Therefore, in order to determine the precise precursor-product relationships for these compounds, enzymatic studies are necessary. As a first step in this direction, Hahlbrock *et al.* (1968) have partially purified a glucosyl transferase from acetone powders of flax seedlings that will catalyse the formation of linamarin from UDP-glucose and acetone cyanohydrin. It is to be hoped that other enzymes catalysing reactions shown in Fig. 6 can be detected.

Finally, it should be noted that the partial biosynthetic pathway presented in Fig. 6 differs greatly from one that recently appeared (Tschiersch, 1967) in which amides were suggested as precursors. This scheme was proposed by Tschiersch (1966b) when the corresponding amines and β-hydroxyamino acids were found not to act as precursors of the cyanogenic glucosides. When isobutyramide-U-^{14}C, (Tapper, unpublished) α-hydroxyisobutyramide-U-^{14}C and the glucoside of α-hydroxyisobutyramide-U-^{14}C (Butler and Conn, 1964b) were administered to flax shoots, no significant labelling of linamarin was observed as compared with valine-U-^{14}C as a precursor. This direct experimental evidence would therefore rule out the scheme proposed by Tschiersch.

IV. BIOSYNTHESIS OF MUSTARD OIL GLUCOSIDES

It is relevant at this point to consider some of the recent studies on the biosynthesis of the mustard oil glucosides or glucosinolate compounds, two examples of which are isopropyl glucosinolate (III) and benzyl glucosinolate (IV). The earlier biosynthetic studies on this interesting group of compounds have recently been reviewed (Underhill and Wetter, 1966).

As proposed by Kjaer (1960), amino acids are converted to aglycones of the thioglucosides in the intact plant. Thus, Benn (1962) and Underhill *et al.* (1962) independently reported the conversion of phenylalanine to glucotropaeolin

Isopropyl
glucosinolate

(III)

Benzyl
glucosinolate

(IV)

(benzyl glucosinolate IV) in *Tropaeolum majus* L. Underhill *et al.* (1962) went on to show that gluconasturtiin, the next higher homologue of glucotropaeolin, was derived from γ-phenylbutyrine (2-amino-4 phenylbutyric acid) in watercress (*Nasturtium officinale* R. Br.). Phenylbutyrine can apparently be formed from phenylalanine by a chain lengthening process involving acetate (Underhill, 1965).

In order to learn more about the intermediates involved as the amino acid is converted to the glucosinolate, Underhill and Chisholm (1964) investigated the source of the thioglucoside nitrogen. They found that doubly labelled L-phenylalanine-U-^{14}C-^{15}N was converted to benzylglucosinolate without a change in the ^{14}C-^{15}N ratio. This observation made it necessary that all intermediates between the amino acid and the thioglucoside be nitrogenous and therefore similar to the situation in cyanogenic glucoside biosynthesis (Butler and Conn, 1964a). When, however, five other nitrogenous compounds including phenylacethydroxamic acid-1-^{14}C and phenylpyruvic acid oxime-2-^{14}C were tried, none served effectively as precursor. While the nitrogen atom in gluconasturtiin has also been found to arise from the amino nitrogen of L-γ-phenylbutyrine-2-^{14}C-^{15}N (Underhill, 1965), there was no definite indication of the nature of the nitrogenous intermediates at the time that Underhill and Wetter (1966) prepared their review.

After isobutyraldoxime-U-^{14}C was shown to act as a precursor of linamarin in flax seedlings, Tapper and Butler (1967) promptly tested whether the same aldoxime could be converted to isopropyl glucosinolate (glucoputranjivin III) in *Cochlearia officinalis* L. Their data, reproduced in Table X, shows that the aldoxime was about ten times as effectively incorporated into the isopropyl-glucosinolate as was L-valine-U-^{14}C. While the aldoxime was readily converted, the oxime of the corresponding keto acid, α-ketoisovaleric acid oxime-U-^{14}C was not. These results were, therefore, in agreement with the earlier findings of Underhill and Chisholm (1964) who had also tried the oxime of an α-keto acid with no success. Tapper and Butler then went on to show that phenylacet-aldoxime-U-^{14}C was about six times more effectively converted to benzyl glucosinolate (IV) in *Lepidium sativum* L. than was phenylalanine-U-^{14}C. Underhill (1967) then independently reported that the same aldoxime was as good a precursor of benzyl glucosinolate as DL-phenylalanine-2-^{14}C in *Tropaeolum majus*. Data from these papers are assembled in Table XI. As further evidence that aldoximes are natural intermediates in the biosynthesis of mustard oil glucosides, Underhill (1967) reported that radioactive phenylacetaldoxime was isolated in a trapping experiment from *T. majus* shoots that had been simultaneously administered DL-phenylalanine-2-^{14}C and inactive phenylacetaldoxime.

TABLE X

Comparison of precursors of isopropylglucosinolate

	Compound administered	Amount (μmole)	Sp. act. (μC/mmole)	% ^{14}C converted
Expt. 1	L-Valine-U-^{14}C[a]	1·0	2490	0·87
	α-Ketoisovaleric acid[a] oxime-U-^{14}C	1·2	1150	< 0·25
Expt. 2	L-Valine-U-^{14}C[a]	4·0	622	0·70
	L-Valine-U-^{14}C[a]	1·0	2490	1·74
	Isobutyraldoxime-U-^{14}C	3·8	390	15·9

Data from Tapper and Butler (1967).
[a] Corrected for assumed loss of carboxyl carbon atom.

TABLE XI

Comparison of precursors of benzylglucosinolate

Expt.	Compound administered	Amount (μmole)	Sp. act. (μC/mmole)	Dilution factor	% ^{14}C converted
BT[a]	L-Phenylalanine-U-^{14}C[c]	1·1	440	110	4·3
	Phenylacetaldoxime-U-^{14}C	1·2	23·1	19	26
EU 1[b]	DL-Phenylalanine-2-^{14}C	48	167	21	11·3
	Phenylacetaldoxime-1-^{14}C	43	32	22	15·1
EU 2[b]	DL-Phenylalanine-2-^{14}C	46	176	29	10·8
	Phenylacetaldoxime-1-^{14}C	40	32	20	12·9

[a] Data from Tapper and Butler (1967).
[b] Data from Underhill (1967).
[c] Corrected for assumed loss of carboxyl carbon atom.

These recent results showing that aldoximes may be intermediates in the formation of both the cyanogenic glycosides and the mustard oil glucosides suggest a common biosynthetic route for these two types of secondary plant products. While the corresponding α-ketoximes might also be expected to be precursors of both types of compounds, the failure of ketoximes to be converted to the glucosinolates places their position in a common scheme in doubt. For that matter the α-ketoximes may serve as precursors of the cyanogenic glucosides only because the nitriles, known to be easily formed from the ketoximes (Ahmad and Spenser, 1961), are in turn efficiently converted to the cyanogens.

Although no information is available on the formation of α-ketoximes from the precursor amino acids, it is tempting to postulate that an α-ketoxime may be formed by oxidation first of the amino acid to the corresponding N-hydroxy-amino acid, and then a second oxidation would yield the ketoxime (Fig. 7). Spenser and Ahmad (1961) have reported that N-hydroxyamino acids undergo a disproportionation reaction to yield one mole each of amino acid and ketoxime. These compounds may therefore be interconverted through oxidation-reduction reactions.

Some evidence for a biological role for an N-hydroxyamino acid is found in recent work on the biosynthesis of the antibiotic hadacidin (Stevens and Emery, 1966). The data indicate that this antibiotic is derived from glycine after initial oxidation to N-hydroxyglycine. Underhill (1967) has also reported preliminary work by H. Kindl on enzyme systems in N. officinale and T. majus which convert N-hydroxyphenylalanine to phenylacetaldoxime. This reaction, if it is confirmed, is an oxidative decarboxylation (2 electrons) in which the nitrogen atom of the amino acid is retained. Such a reaction is similar to the

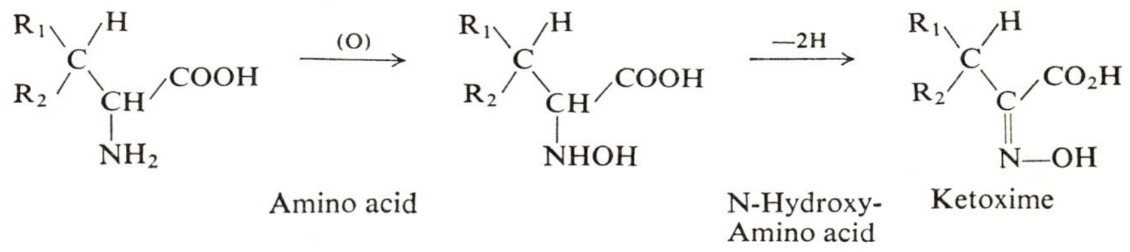

FIG. 7. The oxidation of an amino acid to an α-keto acid oxime.

oxidative decarboxylation (4 electrons) of amino acids that yield the corresponding amide (Mazelis and Ingraham, 1962; Kosuge et al. 1966). Only further study can establish whether a common biosynthetic route is shared between the cyanogenic glucosides and the mustard oil glucosides.

V. METABOLISM OF CYANIDE

Reference has been made (Koukol et al. 1962) to experiments designed to determine whether HCN-^{14}C administered to sorghum seedlings would be converted to dhurrin, the cyanogenic glucoside in that species. Blumenthal-Goldschmidt et al. (1963) made the surprising observation that although dhurrin was not labelled, the HCN-^{14}C was extensively and specifically converted to the amide carbon atom of asparagine. Similar results were obtained when HCN-^{14}C was administered to flax and white clover seedlings, plants also known to contain cyanogenic glucosides. That this was not a unique property of cyanophoric plants was established when barley, pea and red clover seedlings, non-cyanophoric plants, exhibited the same extensive incorporation of HCN-^{14}C into asparagine. Tschiersch (1963) independently made the same observation. Only one plant, the common vetch, *Vicia sativa* L. did not convert HCN-^{14}C into asparagine but instead metabolized the radioactive cyanide to another unknown ninhydrin reacting compound.

In attempting to explain these results at the enzymatic level, we (Blumenthal-Goldschmidt *et al.* 1963) proposed that the enzyme *serine sulfhydrase* (Schlossmann and Lynen, 1957) might react in a non-specific manner with cyanide rather than sulfide to yield an organic nitrile rather than cysteine. The reaction originally proposed for serine sulfhydrase is shown in Fig. 8. If HCN could substitute for H_2S, the nitrile β-cyanoalanine would be produced.

While this work was in progress, Ressler (1962) who has investigated lathyrism factors in legumes, reported the isolation of β-cyanoalanine from common vetch seed. Extension of our observations permitted Ressler *et al.* (1963) to identify the unknown produced on feeding HCN-^{14}C to common vetch seedlings as the lathyritic dipeptide γ-glutamyl-β-cyanoalanine. Fowden *et al.*

$$+ H_2S \xrightarrow[\text{Sulfhydrase}]{\text{Serine}} \begin{array}{c} CH_2SH \\ | \\ CHNH_2 \\ | \\ CO_2H \\ \text{Cysteine} \end{array} + H_2O$$

$$\begin{array}{c} CH_2OH \\ | \\ CHNH_2 \\ | \\ CO_2H \\ \text{Serine} \end{array}$$

$$+ HCN \xrightarrow{?} \begin{array}{c} CH_2CN \\ | \\ CHNH_2 \\ | \\ CO_2H \end{array} + H_2O$$

$$\beta\text{-Cyanoalanine}$$

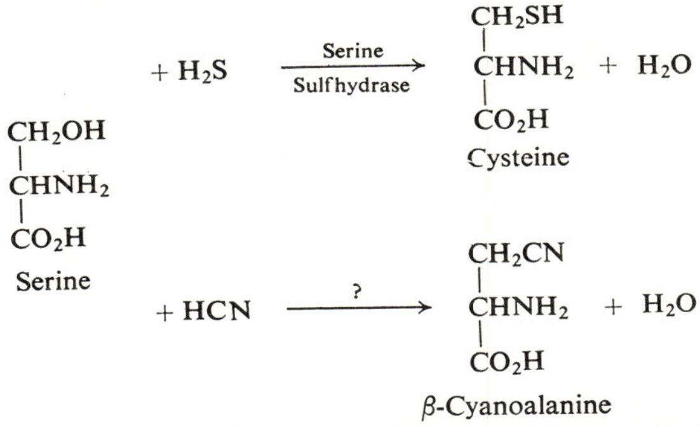

FIG. 8. The reaction catalysed by serine sulfhydrase and the possible formation of β-cyanoalanine.

(1964) independently made the same observation and these findings provided the first indirect evidence that β-cyanoalanine was an intermediate in the assimilation of HCN in plants.

The ability to assimilate HCN is not limited to higher plants but is also exhibited by such familiar organisms as *Escherichia coli* (Dunnill and Fowden, 1965) and *Chlorella pyrenoidosa* (Fowden and Bell, 1965). When the latter organism was administered HCN-^{14}C, both β-cyanoalanine and γ-glutamyl-β-cyanoalanine became labelled. When the kinetics for incorporation of carbon-14 into the two compounds were compared, β-cyanoalanine became labelled first and carbon-14 then moved into the dipeptide. Thus, Fowden and Bell established that β-cyanoalanine was the primary assimilation product.

Fowden and Bell (1965) then went on to observe that the ability of *Vicia sativa* to synthesize the dipeptide rather than asparagine from HCN-^{14}C was due to the presence in that species, as well as several other vetches, of an active γ-glutamyl transferase which catalyses the formation of the dipeptide. Bell and Tirimanna (1965) have reported that one sub-group of vetches possess this dipeptide as a distinguishing characteristic. The other vetches which do not accumulate the dipeptide and instead convert cyanide extensively to asparagine have been shown to lack the glutamyl transferase. These relationships, which are shown in Fig. 9, have been recently reviewed by Fowden (1965).

The extent to which young plant seedlings can metabolize HCN is impressive. In a typical experiment, 1 μmole of HCN can be released as a gas in a volume of 1 litre to give a concentration of about 20 parts per million. If 1 gram of young sorghum seedlings are placed in the container, 15 to 50% of the HCN so administered will be found in asparagine in the plant. The magnitude of this conversion therefore stimulated work on the enzyme system responsible.

In 1965, three laboratories reported preliminary studies on enzymes that catalyse a reaction between HCN and serine. Homogenates of *Lotus tenuis* L.

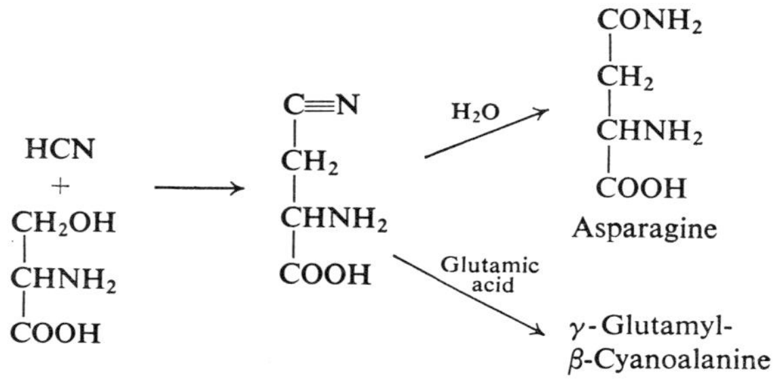

FIG. 9. The formation of asparagine or γ-glutamyl-β-cyanoalanine from β-cyanoalanine.

(Floss *et al.* 1965) were shown to catalyse the formation of β-cyanoalanine from serine and HCN (Reaction 1). However, a partially purified preparation

$$\text{Serine} + \text{HCN} \rightarrow \beta\text{-cyanoalanine} + H_2O \qquad (1)$$

from the same species utilized cysteine much more readily as a substrate. Dunnill and Fowden (1965) observed the enzymatic formation of β-cyano-alanine from serine and HCN in extracts from *E. coli* and found that ATP enhanced the reaction. Tschiersch (1965) also reported an enzymatic reaction between HCN and serine in extracts of *V. sativa*.

The blue lupin (*Lupinus angustifolius* L.) was then examined as an enzyme source because of the large amount of asparagine formed in the etiolated seedlings (Vickery and Pucher, 1943). Hendrickson (1968) has purified an enzyme several hundred-fold from acetone powders of lupin seedlings that requires cysteine as the three carbon substrate and will not utilize serine. The enzyme, β-cyanoalanine synthase (L-cysteine hydrogen sulfide lyase (adding HCN) E.C. 4.4.1._), catalyses Reaction 2 and is conveniently followed by measuring colorimetrically the H_2S formed.

$$\text{cysteine} + \text{HCN} \rightarrow \beta\text{-cyanoalanine} + H_2S \qquad (2)$$

The extensively purified enzyme (Hendrickson and Conn, 1969) will also utilize *O*-acetylserine as the 3-carbon substrate for β-cyanoalanine formation.

This finding is probably related to recent studies on cysteine synthesis in microorganisms (Kredich and Tomkins, 1966) and higher plants (Giovanelli and Mudd, 1967) in which O-acetylserine rather than serine is the preferred substrate for the enzyme cysteine synthase (O-acetylserine sulfhydrase) (Reaction 3).

$$O\text{-acetylserine} + H_2S \rightarrow \text{cysteine} + \text{acetate} \qquad (3)$$

The blue lupin seedling is a rich source of the enzyme cysteine synthase as well as the β-cyanoalanine synthase. The two enzymes are readily separated by differential centrifugation, however, because the latter enzyme is localized on particles (mitochondria) while the cysteine synthase is a soluble protein. The ability of the purified blue lupin β-cyanoalanine synthase to utilize O-acetylserine as well as cysteine may be due to contamination with the cysteine synthase enzyme.

Studies in cyanide metabolism have been extended to fungi by Strobel and his associates. An unidentified psychrophilic basidiomycete (the "snow mold" of alfalfa and grasses) is known to produce HCN in sufficiently high concentrations to poison the host plant (Ward and Lebeau, 1962). Strobel (1964), in questioning how the fungus could survive when the host plant could not, administered HCN-^{14}C to the fungus and found that two amino acids, alanine and glutamic acid, were significantly labelled. Degradation of the alanine showed that the carbon-14 was almost exclusively located in the carboxyl carbon while in the glutamic acid all of the radioactivity was in the alpha carboxyl group. While unknowns in the organic and neutral fractions extracted from the fungus were also extensively labelled, in contrast to the results with higher plants, Strobel (1966) concentrated on the conversion of HCN to the two amino acids. α-Aminopropionitrile labelled exclusively in the nitrile carbon could be isolated from the fungus when it was exposed to HCN-^{14}C, and kinetic studies indicated that the nitrile was a precursor of the alanine that is subsequently formed. Strobel showed that the snow mold contained an enzyme capable of synthesizing α-aminopropionitrile from acetaldehyde, HCN and ammonia, the Strecker synthesis for a nitrile (Reaction 4).

$$CH_3CHO + HCN + NH_3 \rightarrow CH_3CHNH_2CN + H_2O \qquad (4)$$

The fungus apparently also contains an enzyme that catalyses the hydrolysis of the nitrile to form alanine.

In a second study Strobel (1967) showed that the snow mold could also form 4-amino-4-cyanobutyric acid by a similar reaction involving HCN, ammonia and succinic semialdehyde. The use of unlabelled NH_3, $H^{13}C^{15}N$ and succinic semialdehyde-U-^{14}C showed that the nitrile group in the butyric acid derivative came entirely from $H^{13}C^{15}N$ while the amino group was derived exclusively from the unlabelled NH_3. Again evidence was provided for enzymes that catalyse this second Strecker synthesis as well as the hydrolysis of the nitrile to glutamic acid. Finally, a metabolic cycle involving succinic semialdehyde, cyanobutyric acid, glutamic acid and γ-aminobutyric acid and accomplishing the conversion of HCN to CO_2 and NH_3 was proposed as a detoxification mechanism for cyanide.

Other fungi have been examined for their ability to assimilate HCN-[14]C (Allen and Strobel, 1966) and several like the snow mold converted the carbon-14 in the cyanide to alanine. Only one, a *Fusarium*, converted HCN-^{14}C to asparagine as do higher plants, and in several, including another *Fusarium*, HCN-^{14}C was not converted to any labelled amino acid.

Several of the fungi which assimilate HCN are also known to produce cyanide. This includes the snow mold, and the origin of the HCN in the fungus has been studied by Ward and his associates. A recent publication (Ward, 1964) reports that a water soluble, heat-labile compound that decomposes rapidly at pH values above 6 is the cyanogenic substance. Subsequently, Ward and Thorn (1966) presented experiments showing that glycine promoted the formation of this labile cyanogenic substance, and that both HCN and the intermediate cyanogen were labelled when glycine-2[14]C was administered to the fungus. On the other hand, Stevens and Strobel (1968) have provided convincing evidence that linamarin and lotaustralin are produced by the snow mold, that these cyanogenic glucosides are formed from valine and isoleucine, and that enzymes catalysing the hydrolysis of the glucosides and dissociation of the acetone cyanohydrin are present in the fungal tissues. The available evidence therefore favours the view that cyanogenesis in the fungus parallels the phenomenon as it occurs in higher plants, i.e. cyanogenic glucosides are the source of the HCN. HCN assimilation in the same organism, however, obviously does not follow the route employed by plants.

While cyanide formation in bacteria has been known for many years, it is only recently that the origin of the HCN has been examined. Michaels *et al.* (1965) have shown that the nitrile carbon of the HCN produced by *Chromobacterium violaceum* is rather directly derived from the alpha carbon atom of glycine. While cyanoformic acid has been proposed as an intermediate, it is also possible to write reactions in which the oximes of glyoxylic acid and formaldehyde are intermediates. Such a scheme would be in agreement with oxime intermediates in cyanogenesis in higher plants, but no direct experimental evidence is available on this point. *C. violaceum* used in these studies has also been shown to assimilate HCN and the first assimilation product appears to be β-cyanoalanine (Brysk *et al.* 1968).

VI. CONCLUSION

These studies on the metabolism of cyanogenic glucosides and related compounds have been rewarding in that they have suggested other problems, some of which have yet to be investigated. The precursor-product relationship between α-amino acid and aglycone has been well established for prunasin, dhurrin, taxiphyllin, linamarin and lotaustralin. It will be interesting therefore to see whether *Zieria laevigata* produces zierin (*m*-hydroxymandelonitrile-β-D-glucoside) (Finnemore and Cooper, 1936) from *m*-hydroxyphenylalanine or whether prunasin is converted to zierin by *meta*hydroxylation. Similarly acacipetalin (β-dimethyl-α-hydroxyacrylonitrile-β-D-glucoside) (Rimington, 1935) should be formed from leucine.

The precursor-product relationship between amino acids and the cyanogens caused Abrol and Conn (1966) to re-examine the nature of the cyanogenic material in *Lotus arabicus* L. seedlings. The cyanogen had been reported to be *lotusin*, a flavonoid joined to maltose cyanohydrin in ether linkage (Dunstan and Henry, 1901). Since no known amino acid would give rise to a glucose cyanohydrin by the scheme represented in Fig. 6, it seemed likely that the structure of the cyanogenic material was in error. Careful analysis of the *Lotus* seedlings showed that linamarin and lotaustralin were the only cyanogens that were readily detected. Moreover, these two glucosides accounted for all of the HCN that was produced in the seedlings.

The extensive metabolism of HCN by both cyanophoric and non-cyanophoric plants obviously raises the question of the significance of this assimilation process. In the cyanophoric plant, the conversion of HCN to asparagine clearly might be a detoxification mechanism. Abrol and Conn (1966) showed that the radioactivity in valine-^{14}C and isoleucine-^{14}C fed to *Lotus* seedlings first was incorporated into linamarin and lotaustralin in those seedlings. In the intact plants however, the cyanogenic glucosides were broken down at a significant rate and the HCN-^{14}C produced was then incorporated into asparagine. Thus asparagine-^{14}C with a major portion of its radioactivity in the amide carbon was obtained from plants fed L-valine-U-^{14}C. Abrol *et al.* (1966) were also able to show that L-tyrosine, appropriately labelled with carbon-14, gave rise to the cyanogenic glucoside in *Nandina domestica* Thunb. and asymmetrically labelled the asparagine as well. Here again an active turnover of the cyanogen was indicated.

A different role for cyanide assimilation can be proposed for those plants containing β-cyanoalanine and its derivatives. Since asparagine does not appear to be converted to β-cyanoalanine, a cyanogenic glucoside would constitute a good alternate source of HCN and perhaps the only source of carbon for the C_4-position of β-cyanoalanine. Tschiersch (1966a) could show that phenylalanine-2-^{14}C fed to *Vicia angustifolia* seedlings labelled β-cyanoalanine in the nitrile carbon predominantly. This species contains the cyanogen vicianin which would then be functioning as an intermediate in the pathway between phenylalanine and β-cyanoalanine. When certain other plants (e.g. *Lathyrus* species) containing lathyrism factors are considered, the situation is complicated by the fact that the genus *Lathyrus* apparently lacks cyanogenic glucosides.

Of even greater interest to the plant biochemist is the possible significance of cyanide assimilation in the biosynthesis of asparagine. While the biosynthesis of glutamine in plants is known to occur from glutamic acid, ammonia and

ATP in the presence of the enzyme *glutamine synthetase*, no comparable, well-characterized enzyme for asparagine biosynthesis has been detected in higher plants (Fowden, 1967). Any significance of cyanide metabolism in asparagine biosynthesis is, of course, dependent upon a source of cyanide. While the cyanogenic glucosides would constitute prime candidates for this role, they are not by any means ubiquitous in their natural occurrence. This fact can always be countered with the argument that cyanogenic glucosides may be present in many plants but at such low concentrations as to have escaped detection by the relatively insensitive tests that investigators have used.

To test this point we have, in the specific case of blue lupin, administered as carbon-14 labelled compounds the four amino acids phenylalanine, tyrosine, valine and isoleucine known to give rise to cyanogenic glycosides. We then investigated the ^{14}C-labelling in the asparagine formed in these experiments. While the amide was labelled, the asymmetric distribution of carbon-14 to be expected if cyanide were an intermediate was not found. Similar results were obtained with glycine-2-^{14}C.

These last observations suggest that the role for cyanide assimilation is one of detoxification that is useful in the case of cyanophoric plants. This may be a metabolic activity acquired early in evolution and retained by species that no longer have a need for such a process. The similarity to one of the arguments presented for the "raison d'être" of secondary plant products (Fraenkel, 1959) is intriguing.

ACKNOWLEDGEMENTS

Supported in part by Grant GM-5301 from the Institute for General Medical Sciences, National Institutes of Health, U.S. Public Health Service.

REFERENCES

Abrol, Y. P., and Conn, E. E. (1966). *Phytochemistry*, **5**, 237
Abrol, Y. P., Conn, E. E., and Stoker, J. R. (1966). *Phytochemistry*, **5**, 1021.
Ahmad, A., and Spenser, I. D. (1961). *Can. J. Chem.* **29**, 1340.
Akazawa, T., Miljanich, P., and Conn, E. E. (1960). *Plant Physiol.* **35**, 535.
Albers, H. (1941). *In* "Die Methoden der Ferment Forschung" (E. Bamann and K. Myrbäck, eds.), Vol. 3, pp. 2137–2149, George Thieme, Leipzig.
Allen, J., and Strobel, G. A. (1966). *Can. J. Microbiol.* **12**, 414.
Becker, W., Benthin, U., Eschenhof, E., and Pfeil, E. (1963). *Biochem. Zeit.* **337**, 156.
Bell, E. A., and Tirimanna, A. S. L. (1965). *Biochem. J.* **97**, 104.
Benn, M. H. (1962). *Chem. Ind.* 1907.
Ben-Yehoshua, S., and Conn, E. E. (1964). *Plant Physiol.* **39**, 331.
Bleichert, E. F., Neish, A.C., and Towers,G. H. N. (1966). *In* "Biosynthesis of Aromatic Compounds", Proceedings of the 2nd Meeting of the Federation of European Biochemical Societies (G. Billek, ed.), Vol. 3, pp. 119–127, Pergamon Press, Oxford.
Blumenthal-Goldschmidt, S., Butler, G. W., and Conn, E. E. (1963). *Nature, Lond.* **197**, 718.

Bruns, F. H., and Fiedler, L. (1958). *Biochem. Zeit.* **330**, 324.
Brysk, M. M., Corpe,W. A., and Hankes, L. V. (1968). *Bact. Proc.* 115 (Abstract).
Butler, G. W. (1965). *Phytochemistry*, **4**, 127.
Butler, G. W., and Butler, B. G. (1960). *Nature, Lond.* **187**, 780.
Butler, G. W., and Conn, E. E. (1964a). *J. Biol. Chem.* **239**, 1674.
Butler, G. W., and Conn, E. E. (1964b), Tenth International Botanical Congress, Abstracts, No. 375, Edinburgh, p. 164.
Butler, G. W., Bailey, R. W., and Kennedy, L. D. (1965). *Phytochemistry*, **4**, 369
Coburn, R. A., and Long, L., Jr. (1966). *J. org. Chem.* **31**, 4312.
Conn, E. E., and Akazawa, T. (1958). *Fed. Proc.* **17**, 205.
Dilleman, G. (1958). *In* "Handbuch der Pflanzenphysiologie" (W. Ruhland, ed.), Vol. VIII, 1050–1075, Springer, Berlin.
Dunnill, P. M., and Fowden, L. (1965). *Nature, Lond.* **208**, 1206.
Dunstan, W. R., and Henry, T. A. (1901). *Phil. Trans. R. Soc. Lond.* B. **194**, 515.
Finnemore, H., and Cooper, J. M. (1936). *J. Proc. R. Soc. N.S.W.* **70**, 175.
Floss, H. G., Hadwiger, L., and Conn, E. E. (1965). *Nature, Lond.* **208**, 1207.
Fowden, L. (1965). *In* "Biosynthetic Pathways in Higher Plants" (J. B. Prıdham and T. Swain, eds.), pp. 73–99, Academic Press, London.
Fowden, L., and Bell, E. A. (1965). *Nature, Lond.* **206**, 110.
Fowden, L., Conn, E. E., Bell, E. A., and Tirimanna, A. S. L. (1964). *In* "Proceedings of the 1st Meeting of the Federation of European Biochemical Societies", A70, 55, London.
Fowden, L. (1967). *Ann. Rev. Plant Physiol.* **18**, 85.
Fraenkel, G. S. (1959). *Science*, **129**, 1466.
Gander, J. E. (1958). *Fed. Proc.* **17**, 226.
Gander, J. E. (1960). *Plant Physiol.* **35**, 767.
Gander, J. E. (1962). *J. Biol. Chem.* **237**, 3229.
Gander, J. E. (1966). *Phytochemistry*, **5**, 125.
Geissman, T. A., and Hinreiner, E. (1952). *Bot. Rev.* **18**, 77.
Giovanelli, J., and Mudd, S. H. (1967). *Biochem. Biophys. Res. Commun.* **27**, 150.
Hahlbrock, K., Tapper, B. A., Butler, G. W., and Conn, E. E. (1968). *Archs. Biochem. Biophys.* **125**, 1013.
Hegnauer, R. (1963). "Chemotaxonomie der Pflanzen", Vol. 1–4, Birkhauser. Verlag, Basel.
Helferich, B., and Kleinschmidt, T. (1968). *Hoppe-Seyler's Z. phys. Chem.* **349**, 25.
Hendrickson, H. R. (1968). *Fed. Proc.* **27**, 593.
Hendrickson, H. R. and Conn, E. E. (1969). *J. Biol. Chem.* (in press).
Kjaer, A. (1960). *Fort. Chem. Organ. Naturstoffe*, **18**, 122.
Koukol, J., Miljanich, P., and Conn, E. E. (1962). *J. biol. Chem.* **237**, 3223.
Kosuge, T., Heskett, M. G., and Wilson, E. E. (1966). *J. biol. Chem.* **241**, 3738.
Kredich, N. M., and Tomkins, G. M. (1966). *J. biol. Chem.* **241**, 4955.
Kumar, S., and Mahadevan, S. (1963). *Archs Biochem. Biophys.* **103**, 516.
Libby, P. (1964). Unpublished results.
Mahadevan, S. (1963). *Archs Biochem. Biophys.* **100**, 557.
Mao, C.-H., and Anderson, L. (1967). *Phytochemistry*, **6**, 473.
Mazelis, M., and Ingraham, L. L. (1962). *J. biol. Chem.* **237**, 109.
Mentzer, C., and Favre-Bonvin, J. (1961). *C.r. hebd. Acad. Sci., Paris*, **253**, 1072.
Mentzer, C., Favre-Bonvin, J., and Massias, M. (1963). *Bull. Soc. Chim. Biol.* **45**, 749.
Michaels, R., Hankes, L. V., and Corpe, W. A. (1965). *Archs Biochem. Biophys.* **111**, 121.
Neish, A. C. (1964). *In* "Biochemistry of Phenolic Compounds" (J. B. Harborne, ed.), 295–359, Academic Press, London.
Plouvier, V. (1935). *C.r. hebd. Acad. Sci., Paris*, **200**, 1985.

Pridham, J. B., and Saltmarsh, M. J. (1963). *Biochem. J.* **87**, 218.
Ressler, C. (1962). *J. biol. Chem.* **237**, 733.
Ressler, C., Giza, Y.-H., and Nigam, S. N. (1963). *J. Am. chem. Soc.* **85**, 2874.
Rimington, C. (1935). *Onderstepoort J. Vet. Sci.* **5**, 445.
Robinson, M. E. (1930). *Biol. Rev.* **5**, 126.
Rosenthaler, L. (1908). *Biochem. Zeit.* **14**, 238.
Schlossmann, K., and Lynen, F. (1957). *Biochem. Zeit.* **328**, 591.
Schrader, J. C. C. (1803). *Gilbert, Anallen* **13**, 503.
Seely, M. K., Criddle, R. S., and Conn, E. E. (1966). *J. biol. Chem.* **241**, 4457.
Spenser, I. D., and Ahmad, A. (1961). *Proc. chem. Soc.* 375.
Stevens, D. L., and Strobel, G. A. (1968), *J. Bact.* **95**, 1094.
Stevens, R. L., and Emery, T. F., (1966). *Biochemistry*, **5**, 74.
Strobel, G. A. (1964). *Can. J. Biochem.*, **42**, 1637.
Strobel, G. A. (1966). *J. biol. Chem.* **241**, 2618.
Strobel, G. A. (1967). *J. biol. Chem.* **242**, 3265.
Tapper, B. A. and Butler, G. W. (1967). *Archiv. Biochem. Biophys.* **120**, 719.
Tapper, B. A., Conn, E. E., and Butler, G. W. (1967). *Archs Biochem. Biophys.* **119**, 593.
Tschiersch, B. (1963). *Flora*, **153**, 115.
Tschiersch, B. (1965). *Flora, Abt. A.* **156**, 363.
Tschiersch, B. (1966a). *Flora, Abt. A.* **157**, 43.
Tschiersch, B. (1966b). *Flora, Abt. A.* **157**, 358.
Tschiersch, B. (1967). *Die Pharmazie*, **22**, 76.
Underhill, E. W. (1965). *Can. J. Biochem.* **43**, 179.
Underhill, E. W. (1967). *Eur. J. Biochem.* **2**, 61.
Underhill, E. W., and Chisholm, M. D. (1964). *Biochem. Biophys. Res. Commun.* **14**, 425.
Underhill, E. W., and Wetter, L. R. (1966). *In* "Biosynthesis of Aromatic Compounds", Proceedings of the 2nd Meeting of the Federation of European Biochemical Societies (G. Billek, ed.), Vol. 3, 129.
Underhill, E. W., Chisholm, M. D., and Wetter, L. R. (1962). *Can. J. Biochem. Physiol.* **40**, 1505.
Uribe, E. (1965). Ph.D. Thesis, University of California at Davis.
Uribe, E., and Conn, E. E. (1966). *J. biol. Chem.* **241**, 92.
Vickery, H. B., and Pucher, G. W. (1943). *J. biol. Chem.* **150**, 197.
Ward, E. W. B. (1964). *Can. J. Bot.* **42**, 319.
Ward, E. W. B., and Lebeau, J. B. (1962). *Can. J. Bot.* **40**, 85.
Ward, E. W. B., and Thorn, G. D. (1966). *Can. J. Bot.* **44**, 95.

CHAPTER 3

Recent Investigations on the Biosynthesis of Carotenoids and Triterpenes

T. W. GOODWIN

Biochemistry Department, University of Liverpool, Liverpool, England

I. Introduction	75
II. Stereospecific Biosynthesis of Squalene and Phytoene	75
III. Mechanism of Formation of Cyclic Carotenes	78
IV. Formation of Phytoene and its Subsequent Desaturation . . .	82
V. Phytoene Synthesis by Isolated Chloroplasts	84
VI. The Mechanism of Formation of Cycloartenol	85
Acknowledgements	89
References	89

I. Introduction

General reviews of carotenoid biosynthesis have been recently presented by Goodwin (1965) and more recently by Porter and Anderson (1967). It is, thus, only necessary to outline some of the more recent developments in the study of carotenoid and triterpene biosynthesis and, in order to highlight certain problems, consideration is given mainly to investigations carried out in the author's laboratory, first at Aberystwyth and later at Liverpool, concerned with information obtained with the help of various species of stereospecific labelled mevalonic acid. This approach was introduced by Popják and Cornforth (1966) to study the stereochemical details of cholesterol biosynthesis in animals, and it represents a major landmark in the development of biochemistry.

II. Stereospecific Biosynthesis of Squalene and Phytoene

The universal biological isoprene precursor is isopentenyl pyrophosphate (I) (IPP) which is formed in three well-authenticated steps from mevalonic acid (II) (Scheme I). The first step of the chain-elongation process involves isomerization of IPP to dimethylallyl pyrophosphate (DMAPP) (III) and the basic mechanism involved is indicated in Scheme II (Agranoff *et al.* 1969). DMAPP

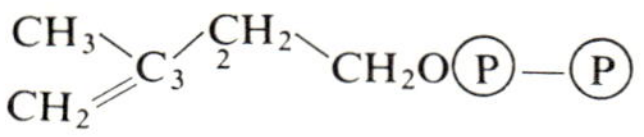

(I) Isopentenyl pyrophosphate

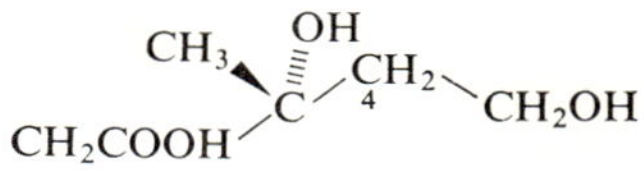

(II) Mevalonic acid

(III) Dimethylallyl pyrophosphate

then acts as starter for chain-elongation and Scheme III indicates how it condenses with a molecule of IPP to form geranyl pyrophosphate (GPP) (C_{10}). A similar condensation of GPP with a further molecule of IPP yields

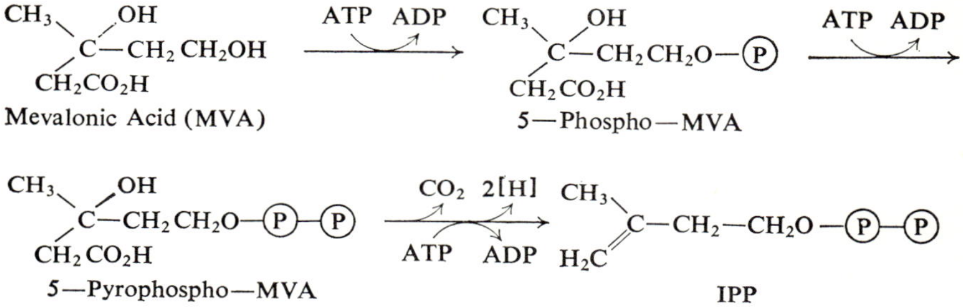

Mevalonic Acid (MVA)

5—Phospho—MVA

5—Pyrophospho—MVA

IPP

Scheme I

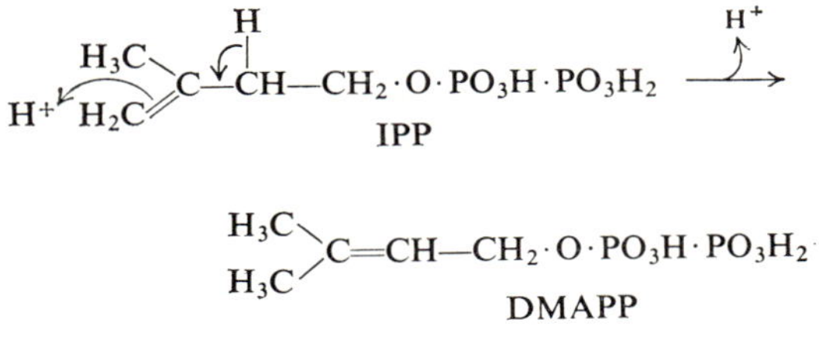

IPP

DMAPP

Scheme II

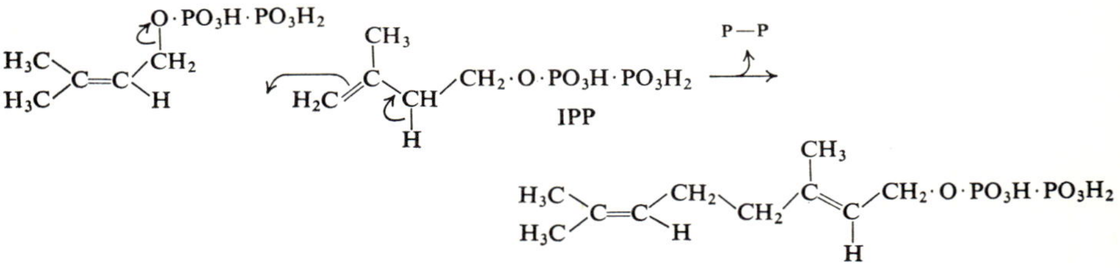

IPP

Scheme III

Geranyl pyrophosphate

farnesyl pyrophosphate (FPP) (C_{15}) which is the precursor of squalene (IV). FPP can then combine with yet another molecule of IPP to give geranylgeranyl pyrophosphate (GGPP) (C_{20}) which is the precursor of phytoene (V). In each of these steps a hydrogen is lost from C-2 of IPP, which originated from C-4

(IV) Squalene

of MVA. Thus, in forming a molecule of squalene from six molecules of MVA, six protons are lost from the six carbon atoms originating from C-4 of MVA.

(V) Phytoene*

* Naturally occurring phytoene has its central double bond in the *cis*-configuration, but for simplication in considering changes not involving this double bond, the structure is drawn with the *trans*-configuration.

In the case of phytoene, eight protons are lost from eight carbon atoms originating from the same source. The C-4 of MVA carries two hydrogen atoms and in the enzymic reactions involved in squalene and phytoene synthesis it is likely that the removal of hydrogen in each step is stereospecific; in purely chemical reactions this would not be so. To test this idea experimentally it was necessary to distinguish between the two hydrogen atoms on C-4 of MVA. This was brilliantly achieved by Cornforth and Popják and their colleagues (Cornforth *et al.* 1965) who synthesized two species of MVA stereospecifically labelled with deuterium or tritium so that C-4 had either the *R*(VI) or *S*(VII)

(VI) 4-*R*-Configuration   (VII) 4-*S*-Configuration

configuration.* When mixed with MVA labelled with ^{14}C at C-2, then the two doubly labelled species [2-^{14}C-4R-4-^{3}H$_1$] MVA and [2-^{14}C-4-*S*-4-^{3}H$_1$] were available for study. If there is stereospecific removal of the *pro-S* hydrogen* from C-4 of MVA at each step during the formation of the squalene then with

* The Cahn and Ingold (1951) nomenclature as extended by Hanson (1966) to naming the paired ligands *g* and *g* at a tetrahedral atom X_{ggij}.

[2-^{14}C-4R-4-^{3}H$_1$] MVA all the tritium would be retained and the ^{14}C:^{3}H ratio in squalene would be the same as in the starting substrate. If, on the other hand, the *pro-R* hydrogen is removed then no tritium will appear in the squalene. With [2-^{14}C-4S-4-^{3}H$_1$] MVA the reverse would be expected. If no stereo-selectivity exists then the ratio would be 1:0·5 in each case; varying stereo-selectivites at different steps would give intermediate values. Cornforth *et al.* (1966) converted [4R-^{2}H$_1$] MVA and [4S-^{2}H$_1$] MVA into farnesyl pyrophosphate by a liver enzyme. In each case farnesol was liberated by alkaline phosphatase and examined in the mass spectrometer. The specimen from the [4S-^{2}H$_1$] MVA was unlabelled whilst that from the [4R-^{2}H$_1$] MVA contained a trideuterated species. This observation was confirmed by experiments on squalene synthesis by a liver system with [2-^{14}C-4R-4-^{3}H$_1$] MVA and [2-^{14}C-4S-4-^{3}H$_1$] MVA as substrates. In the first case all the tritium was retained and in the second all was eliminated; thus in the formation of a molecule of squalene in liver the loss of a proton from C-4 of each of the six molecules of MVA is stereospecific and in the same sense at each step. The *pro-R* hydrogen is retained and the *pro-S* hydrogen is eliminated. Using the tritiated substrates of Cornforth and Popják, we showed that the same situation held for the formation of squalene and for the synthesis of phytoene, the first C-40 carotenoid precursor, in plants and micro-organisms (Goodwin and Williams, 1965; 1966) (Table I).

It should be noted that the double bonds formed in squalene and phytoene with which C-4 of MVA is concerned have the *trans*-configuration, whilst those in rubber have the *cis*-configuration (VIII). In this biosynthesis of rubber, it was found that it was the *pro-S* hydrogen which was retained and the *pro-R* hydrogen which was lost (Archer *et al.* 1966).

(VIII) Rubber

III. MECHANISM OF FORMATION OF CYCLIC CAROTENES

[2-^{14}C-4R-4-^{3}H$_1$] MVA has also been used to investigate the mechanism involved in the formation of the α-ionone and β-ionone rings of the cyclic carotenoids. In the stepwise desaturation of phytoene to neurosporene and lycopene no further protons arising from C-4 of MVA are eliminated (Scheme IV). However in the cyclization step from an unsaturated precursor [there is still doubt as to whether it is lycopene or neurosporene; see e.g. Porter and Anderson, 1967] if a mechanism such as that indicated in Scheme V is functioning in the production of the β-ionone ring, then with [2-^{14}C-4R-4-^{3}H$_1$] MVA as substrate the proton finally eliminated is a tritium ion. So if the ^{14}C:^{3}H atomic ratio is normalized to 8:8 in phytoene, then in β-carotene, which contains two β-ionone residues, it should fall to 8:6, and in α-carotene, with

TABLE I

The ratio of $^3H/^{14}C$ in squalene and phytoene synthesized by isolated carrot root slices from $[2-^{14}C-(4R)-4-^3H_1]$ mevalonic acid (Goodwin and Williams, 1966)

Substance	Assay 1	Assay 2	Assay 3
Mevalonic acid (control)	5·99	6·09	—
Squalene (ex liver)	5·76	5·77	5·59
Squalene			
Sample 1	5·46	5·71	5·53
Sample 2	5·43	5·62	5·30
Phytoene			
Sample 1	5·31	5·61	5·46
Sample 2	5·52	5·32	5·89

only one β-ionone residue, it should be 8:7. This was demonstrated in experiments using carrot root slices (Table II) (Goodwin and Williams, 1965).

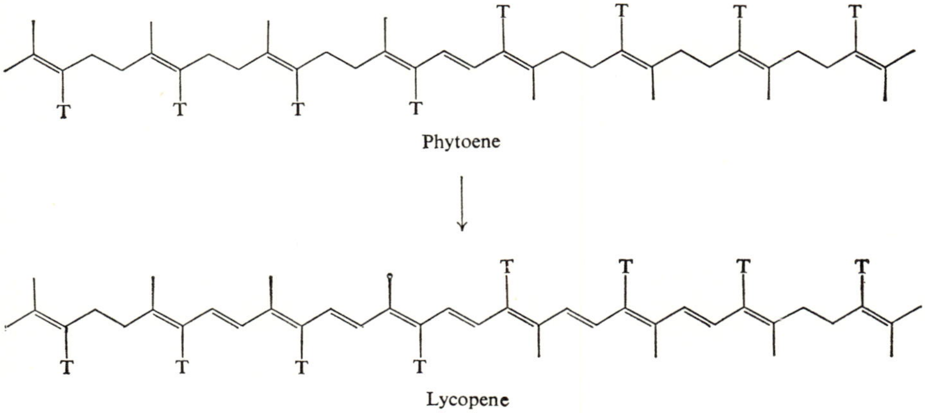

Phytoene

Lycopene

Scheme IV

Confirmatory results were later obtained in experiments on the fruit of *del*-tomatoes, and extended to include γ-carotene (one β-ionone and one open end) δ-carotene (one α-ionone ring and one open end) (Table III) (Williams *et al.* 1967a).

These results excluded the possibility of the α-ionone ring being formed by "flipping over" the double bond in a pre-formed β-ionone ring, because if this were so the atomic ratio in α-carotene would be the same as that in β-carotene, 8:6. However, they do not rule out the possibility that the β-ionone residue is formed from the α-ionone residue. The more likely possibility is that the α-ionone ring is formed from the same carbonium ion as β-carotene, but that it

TABLE II

Comparison of the incorporation of $[2-^{14}C,4R-^3H]$ mevalonic acid into squalene, phytoene, α-carotene and β-carotene in carrot-root slices (Goodwin and Williams, 1965)

Substance	Radioactivity (disintegrations/min)		Mean $^3H/^{14}C$ radioactivity	$^{14}C/^3H$ atomic ratio
	^3H	^{14}C		
$[2-^{14}C,4R-^3H]$ Mevalonic acid	187300	36950	5·08:1	1:1
Squalene (from liver)	30600	5980	5·12:1	6:6·0
Squalene	1000	195	5·13:1	6:6·1
Phytoene	740	144	5·14:1	8:8·1
α-Carotene	470	110	4·27:1	$8:6·7 \pm 0·2^a$
β-Carotene	1470	410	3·59:1	$8:5·7 \pm 0·1^a$

^{a} Standard error of counting.

TABLE III

Incorporation of $(3RS)-[2-^{14}C,(4R)-4-^3H]$ MVA into carotenoid polyenes in *del* tomatoes (Williams *et al.* 1967a)

Polyene	Radioactivity (disintegrations/min)		Mean $^3H/^{14}C$ radioactivity ratio	Mean $^3H/^{14}C$ atomic ratio
	^3H	^{14}C		
$[2-^{14}C-(4R)-4-^3H_1]$ MVA	23910	4450	5·37	1:1
Squalene	328800	65500	5·02	6:6
Phytoene	8470	1700	4·98	8:8
ζ-Carotene	350	70	4·95	8·05 ($\pm 0·31$):8
α-Zeacarotene	111	20	5·07	8·19 ($\pm 0·78$):8
δ-Carotene	35600	7620	4·70	7·59 ($\pm 0·04$):8
ϵ-Carotene	529	105	5·04	8·04 ($\pm 0·33$):8
γ-Carotene	5130	1170	4·38	7·05 ($\pm 0·07$):8
α-Carotene	1770	406	4·36	6·95 ($\pm 0·11$):8
β-Carotene	4470	1205	3·71	5·95 ($\pm 0·33$):8
Lycopene	4640	960	4·84	7·76 ($\pm 0·12$):8

is stabilized by loss of a proton from the carbon atom on the opposite side to that involved in the β-ionone formation (Scheme V). This carbon atom originates from C-2 of MVA, so the problem was tackled with the help of $[2-^{14}C-2-^3H_2]$ MVA. In this case, if the atomic ratio $^{14}C:^3H$ is normalized as 8:16 in phytoene because no protons are lost from C-2 of MVA during its formation, then that of β-carotene would be 8:12 and that of α-carotene 8:11

(Scheme VI) if the mechanism proposed is operating. If the β-ionone residues are formed from α-ionone residues then the ratio for β-carotene would be 8:10. Experiments with *del*-tomatoes (Table IV) showed that the ratio for β-carotene was 8:12 and that it clearly contains one tritium atom more than

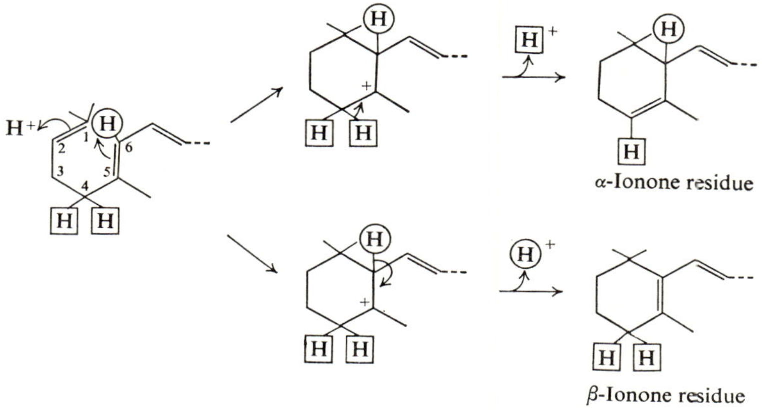

Scheme V. Ⓗ, Labelled from [(4R)-4-^{3}H$_1$]MVA; ⬚H, labelled from [2-^{3}H$_2$]MVA.

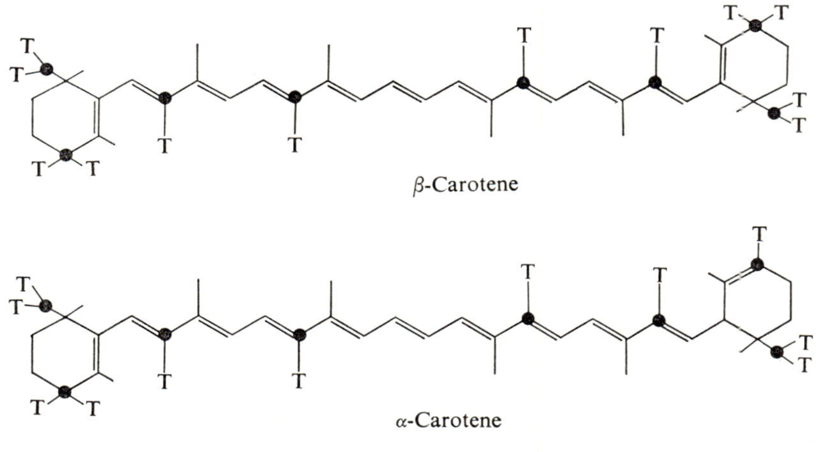

β-Carotene

α-Carotene

Scheme VI (●, Label from [2-^{14}C]MVA; T, label from [2-^{3}H$_2$]MVA.)

α-carotene, which had a ratio of 8:11. Thus it appears that the primary processes in ionone ring formation in the carotenes is the production of a common intermediate, indicated in Scheme V as a carbonium ion, which according to the nature of the active site is stabilized by loss of a proton from different adjacent carbon atoms. A carbonium ion as such is almost certainly not

TABLE IV

Incorporation of $(3RS)$-$[2$-$^{14}C,2$-$^3H_2]$ MVA into squalene, phytoene, β-carotene and α-carotene by slices of carrot root (Williams *et al.* 1967a)

Polyene	Radioactivity (disintegrations/min)		Mean $^3H/^{14}C$ radioactivity	Mean $^3H/^{14}C$ atomic ratio
	^3H	^{14}C		
Squalene	85248	5416	15·74	12:6
Phytoene	87028	5568	15·63	15·90 (± 0·30):8
β-Carotene	55175	4575	12·06	12·25 (± 0·25):8
α-Carotene	23074	2075	11·12	11·39 (± 0·10):8

involved in these reactions, but rather an enzyme-substrate complex (compare Scheme VIII). However further information is required before this can be formally set out with confidence.

IV. FORMATION OF PHYTOENE AND ITS SUBSEQUENT DESATURATION

Further consideration of carotenoid biosynthesis indicates that, in the formation of phytoene, two hydrogens are lost from the two central atoms both of which originated from C-5 of MVA; in addition, one hydrogen originating from C-5 of MVA is lost during each of the four steps involved in converting phytoene into the fully unsaturated carotenoid such as lycopene (Scheme VII). No hydrogen originating from C-5 of MVA is involved in any cyclization reaction. When the stereochemistry of these hydrogen eliminations was studied with $[2$-^{14}C-$5R$-5-$^3H_1]$ MVA and $[2$-^{14}C-5-$^3H_2]$ MVA as substrates, it was found that in the formation of phytoene, two *pro-R* hydrogens are retained and two *pro-S* hydrogens are eliminated at the centre of the molecule. On the other hand, during desaturation of phytoene to lycopene and the cyclic carotenes one *pro-R* hydrogen is lost at each step whilst all the *pro-S* hydrogens are retained (Williams *et al.* 1967b) (Table V).

The results obtained for phytoene suggest a mechanism for its formation and which requires its central double bond to have the *cis*-configuration, which is the configuration of the natural product. In this proposal (Scheme VIII), which is but one of a number of possibilities and which has previously been considered in a slightly different form for squalene (Cornforth *et al.* 1966), the geranylgeranyl residues are connected via a sulphonium ylide, a mechanism which requires the presence of a thio-ether grouping, such as a methionine residue, at the active site of the phytoene synthase. This thio-ether group displaces the pyrophosphate from a molecule of geranylgeranyl pyrophosphate by a S_{N2} substitution reaction which involves inversion of configuration at C-1 of the geranylgeranyl group to give a sulphonium ion. The hydrogens at C-1

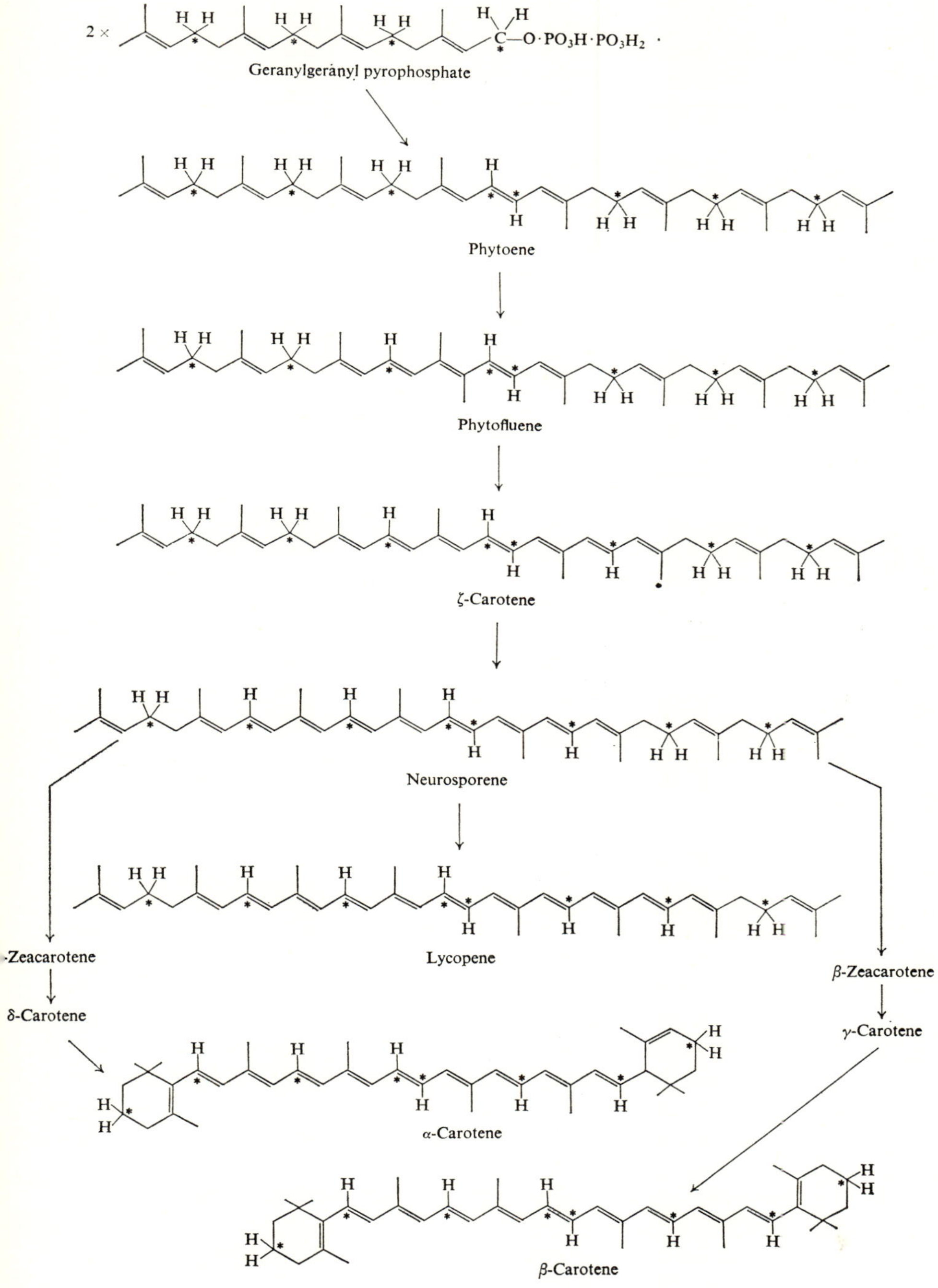
2 ×
O·PO₃H·PO₃H₂
Geranylgeranyl pyrophosphate
Phytoene
Phytofluene
ζ-Carotene
Neurosporene
Zeacarotene
Lycopene
β-Zeacarotene
δ-Carotene
γ-Carotene
α-Carotene
β-Carotene

Scheme VII

TABLE V

Comparison of incorporation of $(3RS)$-$[2$-^{14}C-$(5R)$-5-$^{3}H]$ mevalonic acid into squalene and carotenoid polyenes by slices of *del* tomatoes (Williams *et al.* 1967b)

Substance	Radioactivity disintegrations/min)		Mean $^{3}H/^{14}C$ radioactivity ratio	Mean $^{3}H/^{14}C$ atomic ratio
	^{3}H	^{14}C		
MVA	41830	4061	10·30	
Squalene	281220	24600	11·43	6:6
Phytoene	24380	2056	11·86	8:8
δ-Carotene	28665	4330	6·62	4·46 ($\pm$0·09):8
γ-Carotene	2486	370	6·72	4·54 ($\pm$0·05):8
α-Carotene	4334	673	6·44	4·36 ($\pm$0·11):8
β-Carotene	14985	2345	6·39	4·30 ($\pm$0·06):8
Lycopene	5643	890	6·34	4·23 ($\pm$0·13):8

situated between a double bond and the S atom have a tendency to ionize with the formation of an ylide. This alkylates a second molecule of geranylgeranyl pyrophosphate (again with inversion at C-1) to give a lycopersyl-sulphonium ion (lycopersene is the C_{40} homologue of squalene, see p. 77). The central double bond is then introduced by normal *trans* elimination of the -S- enzyme and a proton from the adjacent methylene group. Retention of both *pro-R*-hydrogens at the centre of the molecule will occur only if the configuration of the newly formed central double bond is *cis*. Furthermore, this mechanism does not require the participation of $NADP^+$ or NAD^+; this has been demonstrated experimentally for isolated chloroplasts (Charlton *et al.* 1967).

V. PHYTOENE SYNTHESIS BY ISOLATED CHLOROPLASTS

Many attempts to prepare by conventional methods chloroplasts which would convert MVA into phytoene were failures, but active preparations were obtained when the chloroplasts were isolated by non-aqueous techniques (Charlton *et al.* 1967). Such preparations, when sonicated will effectively incorporate $[2$-$^{14}C]$ MVA into phytoene in the presence of ATP and Mg^{2+}; no other co-factor is required. A minute amount of squalene is also formed; this is increased in the presence of NADPH, which has no effect on phytoene synthesis. Neither does its presence stimulate the synthesis of any lycopersene (dihydrophytoene). This observation adds further evidence to the conclusion that phytoene and not lycopersene is the first C-40 compound formed in carotenoid biosynthesis (see Goodwin, 1965; Porter and Anderson, 1967 for full details).

We have not yet obtained chloroplasts which will desaturate phytoene to phytofluene, etc. This may be because the method of preparation of the

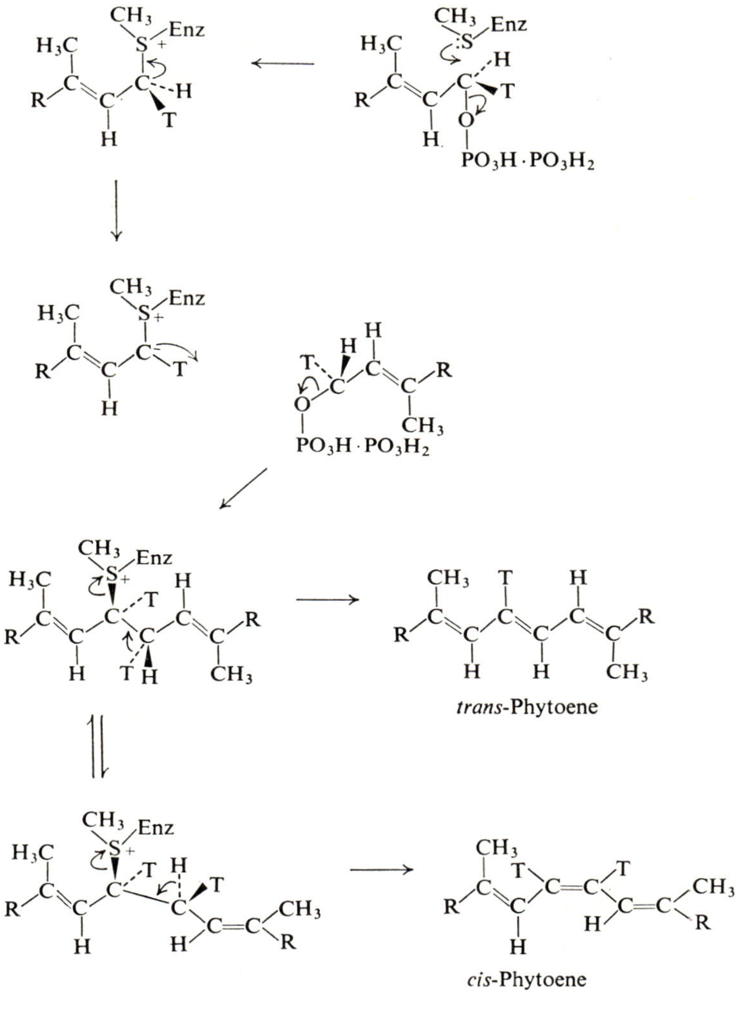

Scheme VIII

chloroplasts removes considerable amounts of lipid material and destroys the structural integrity of the organelle, which is essential for the metabolism of the lipid-soluble phytoene, whilst retaining the enzymes necessary for the formation of geranylgeranyl pyrophosphate and its conversion into phytoene.

VI. The Mechanism of Formation of Cycloartenol

In the biosynthesis of sterols from squalene in animals, the steps are squalene → squalene 2,3-oxide → lanosterol. Squalene oxide has only recently been

4

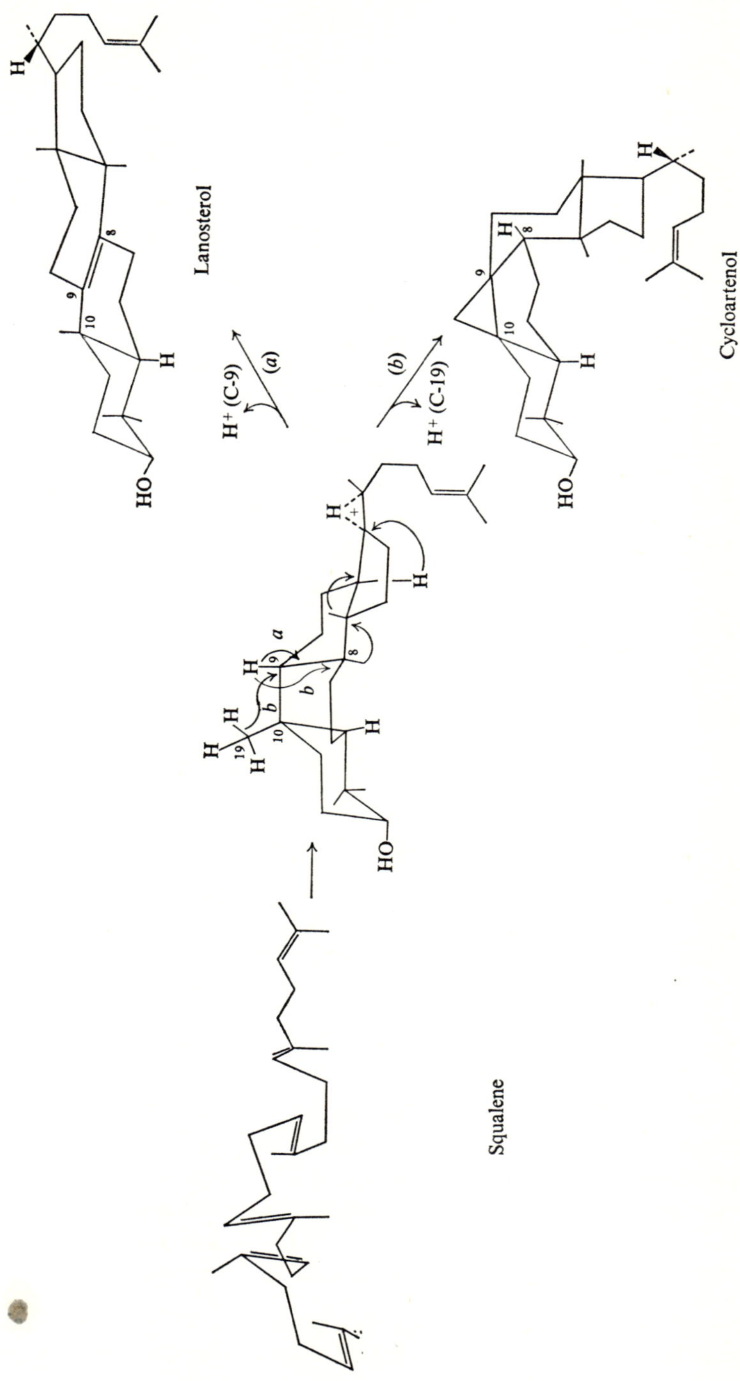

Scheme IX. Cyclization of squalene to lanosterol and cycloartenol.

demonstrated as an intermediate (Corey *et al.* 1966; van Tamelen *et al.* 1966) although lanosterol has been clearly implicated for some time as the first cyclized product (see e.g. Clayton, 1965).

Many recent investigations which employed modern techniques such as thin layer chromatography on silver nitrate, impregnated silica gel plates and GLC separations on the newer columns have indicated the absence of lanosterol from all plants studied, except *Euphorbia* where it is found in the latex; this will be discussed later (see e.g. Goad, 1967). In its place apparently exists

TABLE VI

$^3H/^{14}C$ ratios in the triterpenes isolated from potato leaves after incubation with $[2-^{14}C,(4R)-^3H_1]$-MVA, and in lanosteryl acetate and parkeyl acetates obtained by hydrogen chloride isomerization of the cycloartenyl acetate (Rees *et al.* 1968)

Substance	Radioactivity (counts/min)		$^3H/^{14}C$ radioactivity ratio	$^3H/^{14}C$ atomic ratio[a]
	^3H	^{14}C		
Squalene (sample 1)	35115	5160	6·805	
Squalene (sample 2)	31329	4621	6·780	
Cycloartenyl acetate (isolated by thin-layer chromatography)	11396	1644	6·932	6·12:6
Cycloartenyl acetate (isolated by preparative gas-liquid chromatography)	5752	824	6·981	6·17:6
24-Methylenecycloartanyl acetate	10852	1569	6·917	6·11:6
4-Demethyl phytosterols	8583	1910	4·494	3·31:5
Lanosteryl acetate[b]	5541	1003	5·524	4·88:6
Parkeyl acetate[b]	15132	2243	6·746	5·96:6

[a] Based on the average of the $^3H/^{14}C$ ratios obtained for squalene.
[b] Formed from cycloartenyl acetate on HCl isomerization.

cycloartenol (Scheme IX), which has been detected in trace amounts in all higher plants examined by modern techniques. In the formation of lanosterol from $[2-^{14}C-4R-^3H_1]$ MVA, the ratio of $^{14}C:^3H$ drops from 6:6 to 6:5 because of the ejection of a tritium atom from C-9 in the stabilization of the intermediate by the formation of the double bond at C-8 (Rees *et al.* 1968a) (Scheme IX).

Cycloartenol could be formed either by isomerization of lanosterol or by stabilization of the intermediate by loss of a proton from the methyl group at C-10 and the formation of a cyclopropane ring (Scheme IX). With cycloartenol synthesized from $[2-^{14}C-4R-^3H_1]$ MVA the first possibility would yield

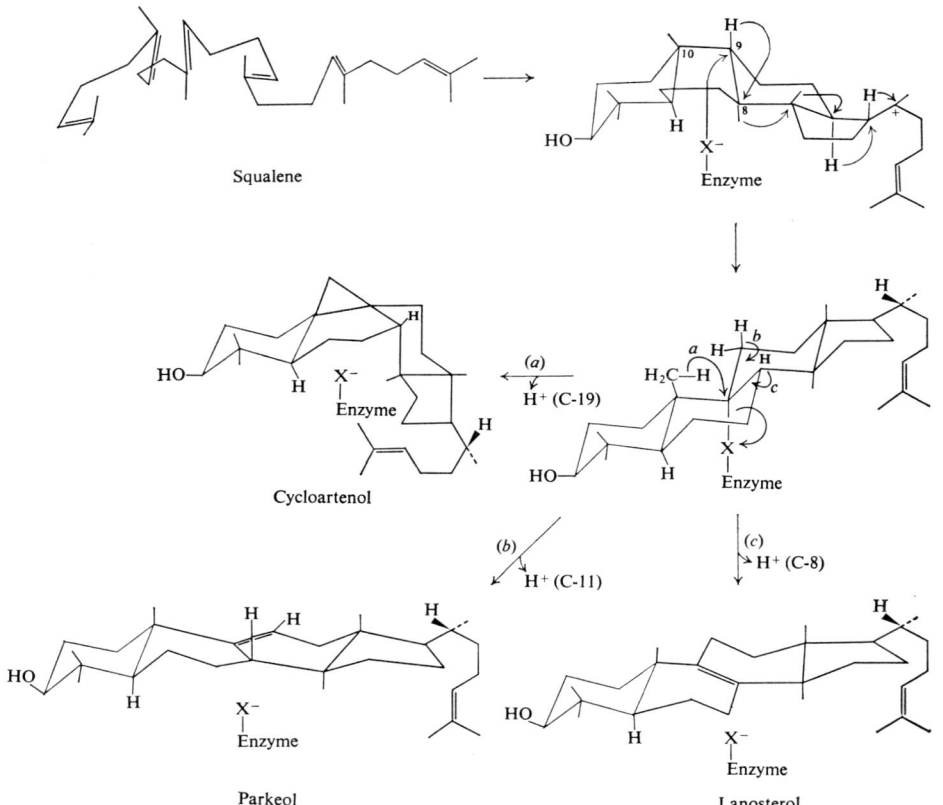

Scheme X. Hypothetical scheme for the enzymic formation of cycloartenol, parkeol and lanosterol.

a product with a $^{14}C:^3H$ ratio the same as that of lanosterol, that is 6:5; on the other hand the second mechanism would yield a ratio of 6:6 because the tritium at C-9 should have migrated to C-8. In experiments with potato leaves, which were used because they accumulate reasonable amounts of cycloartenol (Schreiber and Osske, 1962), a ratio of 6:6 was obtained (Rees *et al.* 1968a). When labelled cycloartenol acetate was isomerized to parkeol and lanosterol (which was in equilibrium with its Δ^7 isomer) it was found that the $^{14}C:^3H$ ratio was 6:6 in parkeol and 6:5 in lanosterol (Table VI). Clearly cycloartenol is a primary product of cyclization and is not formed from lanosterol. The presence of lanosterol in *Euphorbia* latex is due to the presence of an enzyme which will convert cycloartenol into lanosterol (Ponsinet and Ourisson, 1968).

Recently Rees *et al.* (1968b) have obtained a cell-free system from bean leaves which will convert [^{14}C] squalene 2,3-oxide into cycloartenol in good yield without producing any detectable lanosterol.

A mechanism which will explain the formation in Nature of lanosterol, cycloartenol and parkeol is indicated in Scheme XI. If the cyclization product is first stabilized not by proton elimination but by enzyme attack at C-19 as indicated in Scheme X, then the three products can be obtained with the correct stereochemistry by *trans* elimination of the enzyme and a proton from C-8, C-11 and from the methyl at C-10, respectively.

ACKNOWLEDGEMENTS

This work has been supported by generous grants from the S.R.C. and by many able collaborators and colleagues; in particular, the carotenoid investigations have been carried out by Dr G. Britton, Dr R. J. H. Williams and Dr Josephine Charlton and the terpenoid investigations by Dr L. J. Goad and Dr H. H. Rees. The following schemes are reproduced by courtesy of the Biochemical Society: II, III, V, VI (Williams *et al.*, 1967a); VII, VIII (Williams *et al.*, 1967b); IX, X (Rees *et al.*, 1968a).

REFERENCES

Agranoff, B. W., Eggerer, H., Henning, V., and Lynen, F. (1959). *J. Am. chem. Soc.* **81**, 1254.

Archer, B. L., Barnard, D., Cockbain, E. G., Cornforth, J. W., Cornforth, R. H., and Popják, G. (1966). *Proc. R. Soc. B.* **163**, 519.

Charlton, J. M., Treharne, K. J.,and Goodwin, T. W. (1967). *Biochem. J.* **105**, 205.

Clayton, R. B. (1965). *Quart. Rev.* **19**, 168.

Corey, E. J., Russey, W. E., and Ortiz de Montellano, P. R. (1966). *J. Am. chem. Soc.* **88**, 4750.

Cornforth, J. W., Cornforth, R. H., Donninger, C., Popják, G., Ryback, G., Shimizu, Y., Tchii, S., Forchielli, E., and Caspi, E. (1965). *J. Am. chem. Soc.* **87**, 3224.

Cornforth, J. W., Cornforth, R. H., Donninger, C., and Popják, G. (1966). *Proc. R. Soc. B*, **163**, 492.

Goad, L. J. (1967). *In* "Terpenoids in Plants" (J. B. Pridham, ed.) pp. 159–190. Academic Press, London, England.

Goodwin, T. W. (1965). *In* "Biosynthetic Pathways in Higher Plants" (J. B. Pridham and T. Swain, eds.), p. 37, Academic Press, London.

Goodwin, T. W., and Williams, R. J. H. (1965). *Biochem. J.* **97**, 28C.

Goodwin, T. W., and Williams, R. J. H. (1966). *Proc. R. Soc. B*, **163**, 515.

Ponsinet, G., and Ourisson, G. (1968). *Phytochemistry*, **7**, 757.

Popják, G., and Cornforth, J. W. (1966). *Biochem. J.* **101**, 553.

Porter, J. W., and Anderson, D. G. (1967). *Ann. Rev. Plant Physiol.* **18**, 197.

Rees, H. H., Goad, L. J., and Goodwin, T. W. (1968a). *Biochem. J.* **107**, 417.

Rees, H. H., Goad, L. J. and Goodwin, T. W. (1968b).*Tetrahedron Lett.* No. 6, p. 723.

Schreiber, K. and Osske, G. (1962). *Kulturpflanze*, **10**, 372.

Van Tamelen, E. E., Willett, J. D., Clayton, R. B., and Lord, K. E. (1966). *J. Am. chem. Soc.* **88**, 4752.

Williams, R. J. H., Britton, G., and Goodwin, T. W. (1967a). *Biochem. J.* **105**, 99.

Williams, R. J. H., Britton, G., Charlton, J. M., and Goodwin, T. W. (1967b). *Biochem. J.* **104**, 767.

CHAPTER 4

Fatty Acid Biosynthesis in Plants

A. T. JAMES

Unilever Research Laboratories, Sharnbrook, Bedford, England

I. Saturated Fatty Acid Biosynthesis	. . . .	91
II. Unsaturated Fatty Acid Biosynthesis	. . . . .	93
A. Monoenoic Acids	. . . .	93
B. Chain Length Requirements	. . . . . .	98
C. Stereochemistry	. . . .	98
D. 3-*Trans*-hexadecenoic Acid .	. . . . . .	99
E. Polyunsaturated Acids	. . . . . .	100
F. Stereochemistry	. . . .	101
G. Possible Coupling of Fatty Acid and Acyl Lipid Synthesis	. .	101
III. Unusual Fatty Acids	. . . .	102
A. Ricinoleic Acid .	. . . .	102
B. Acetylenic Fatty Acids	. . . . . .	102
IV. The Effect of Temperature on Fatty Acid Biosynthesis	. .	104
References .	. . . .	105

I. Saturated Fatty Acid Biosynthesis

During the period under review, marked advances have been made in the understanding of saturated and unsaturated fatty acid and lipid biosynthesis in higher and lower plants. Earlier work, particularly of Stumpf and his group (Barron *et al.* 1961), was concentrated on the pathways of saturated acid metabolism.

The normal acetate-malonate condensation sequences were demonstrated by Stumpf to operate in seed systems such as the avocado pear, effective substrates being *S*-acetyl and *S*-malonyl Coenzyme A derivatives (Barron *et al.* 1961). However, Vagelos and his coworkers (Alberts *et al.* 1963; Goldman *et al.* 1963a,b; Goldman and Vagelos, 1962; and Lennarz *et al.* 1962) later showed that the heat and acid stable protein fraction required for fatty acid synthesis in *E. coli* was a unique protein of molecular weight 9750, with an attached prosthetic group similar in structure to that of Coenzyme A (I). Like the latter, it contained β-alanine, 2-mercaptoethylamine, pantoic acid and phosphate but it lacked ribose and adenine. The peptide was shown to be bound to 4'-phosphopantetheine, and thus explained the lack of adenine and ribose present in Coenzyme A. It was demonstrated that acyl groups from the

Coenzyme A (I)

acyl-S-CoA thiolesters were transferred to the thiol group of this protein, named the acyl carrier protein or ACP and that this new thiol ester was the true synthetase substrate.

The first steps in the *E. coli* fatty acid synthetase system are thus as follows:

$$\text{Acetyl-}S\text{-CoA} + \text{ACP-SH} \xrightarrow[\text{transacylase}]{\text{acetyl}} \text{acetyl-}S\text{-ACP} + \text{CoA-SH}$$

$$\text{Malonyl-}S\text{-CoA} + \text{ACP-SH} \xrightarrow[\text{transacylase}]{\text{malonyl}} \text{malonyl-}S\text{-ACP} + \text{CoA-SH}$$

From this point in the sequence, the CoA thiol esters are no longer involved and the reactions continue as shown below:

$$\text{Acetyl-}S\text{-ACP} + \text{malonyl-}S\text{-ACP} \xrightarrow[\text{synthetase}]{\beta\text{-ketoacyl-ACP}} \text{acetoacetyl-}S\text{-ACP} + \text{CO}_2 + \text{ACP-SH}$$

$$\text{Acetoacetyl-}S\text{-ACP} + \text{NADPH} \xrightarrow[\text{reductase}]{\beta\text{-ketoacyl-ACP}} \text{D}(-)\text{-}\beta\text{-hydroxybutyryl-}S\text{-ACP} + \text{NADP}$$

$$\text{D}(-)\text{-}\beta\text{-hydroxybutyryl-}S\text{-ACP} \xrightarrow[\text{dehydrase}]{\beta\text{-hydroxylacyl-ACP}} \text{crotonyl-}S\text{-ACP} + \text{H}_2\text{O}$$

$$\text{Crotonyl-}S\text{-ACP} + \text{NADPH} \xrightarrow[\text{reductase}]{\text{enoyl-ACP}} \text{butyryl-}S\text{-ACP} + \text{NADP}$$

The cycle then recommences with butyryl-S-ACP and malonyl-S-ACP until the requisite chain length saturated acid is formed (Vagelos *et al.* 1966).

The β-ketoacyl-ACP synthetase which catalyses the condensation of acetyl-S-ACP and malonyl-S-ACP is known to be inhibited by the sulphydryl reagent iodoacetamide. Alberts *et al.* (1964) showed that this inhibition was abolished by prior incubation with acetyl-S-ACP only. This suggested not only that an active group of the enzyme was an -SH group but also that the acetyl group was transferred from ACP to this group. This acyl carrier protein was soon demonstrated by Stumpf and his group to be involved in plant fatty acid synthesis (Brooks and Stumpf, 1965; Overath and Stumpf, 1964).

Active fatty acid synthesis takes place in all parts of the plant and in recent years leaf and isolated chloroplast systems have been investigated to define their similarity to the seed systems. The initial demonstration of fatty acid synthesis by isolated leaves by Eberhardt and Kates (1957) using $^{14}CO_2$ and ^{14}C-acetate was extended by James (1963) who showed that acetate, octanoate, decanoate and tetradecanoate were utilized by chopped leaves to form longer chain saturated and unsaturated acids. The major site of such synthesis was then found to be the chloroplast (Stumpf and James, 1963), cf. work by Smirnov (1960, 1962, 1964). Although both acetyl-CoA and malonyl-CoA were effectively utilized, it is now known that acetyl-S-ACP and malonyl-ACP are the true substrates (Brooks and Stumpf, 1966). Isolated chloroplasts, like all other synthetase systems, require ATP, Mg^{++}, CO_2, inorganic phosphate and CoA when synthesis is from acetate. In such systems, in addition to the C_{14}, C_{16} and C_{18} saturated straight chain acids, small amounts of the branched and straight chain C_{15} and C_{17} acids are formed (Stumpf and James, 1963).

The effects of light on fatty acid synthesis are still unclear. Stumpf and James (1963) found that synthesis in isolated chloroplasts was greatly diminished in the dark, and inhibited in the light by both NH_3 and PCMU [3-(p-chlorophenyl)1,1-dimethylurea]. Such inhibition could be explained by repression of the photosynthetic production of NADPH and ATP. However, Stumpf et al. (1963) demonstrated a coupling between non-photosynthetic production of NADPH, O_2 and ATP and lipid synthesis, and Stumpf (1967) was unable to replace light by addition of ATP, NADPH and O_2. On the other hand, Mudd and McManus (1962) showed that two fractions could be obtained from disrupted spinach chloroplasts, one of which was soluble and was able to incorporate acetyl-CoA into long chain fatty acids in the dark, provided NADPH and ATP were present. The apparent contradiction in these results has yet to be explained. An excellent review of this field has been given by Stumpf (1967).

II. UNSATURATED FATTY ACID BIOSYNTHESIS

A. MONOENOIC ACIDS

For many years, there was little progress in defining the pathway of formation of unsaturated acids owing to insistence on approaches dependent largely on quantitative analysis. By the use of labelled precursors and refined methods of separation, Bloch and his coworkers were able to define two distinct pathways. In the first system, described originally for yeast (Bloomfield and Bloch, 1960), the Δ^9 unsaturated acids are formed by the direct dehydrogenation of the corresponding saturated acid with CoA, NADPH and molecular oxygen as cofactors. Thus:

$$\text{Stearoyl-}S\text{-CoA} \xrightarrow[\text{NADPH}]{O_2} \text{Oleyl-}S\text{-CoA}$$

4*

In the second system, defined originally with Clostridia (Goldfine and Bloch, 1961), an anaerobic mechanism operated in the following manner:

$$CH_3(CH_2)_8CO_2H \rightarrow CH_3(CH_2)_8CO\text{-}S\text{-}CoA$$

malonyl-S-CoA $\downarrow$

$$CH_3(CH_2)_8CO\cdot CH_2CO\cdot\text{-}S\text{-}CoA + CO_2 + CoA\text{-}SH$$

$\downarrow$

$$\overset{\text{OH}}{\underset{|}{CH_3(CH_2)_8C}}\text{—}CH_2CO\text{-}S\text{-}CoA$$

$\downarrow$

$$CH_3(CH_2)_7CH\overset{cis}{=\!=}CH\text{—}CH_2CO\text{-}S\text{-}CoA$$

$\downarrow$ chain elongation

$$CH_3(CH_2)_7CH\!=\!\!=\!CH\text{—}(CH_2)_7CO_2H$$

Oleic acid.

A generalized scheme proposed by Scheuerbrandt and Bloch (1962) for synthesis of saturated and unsaturated acids by this pathway is shown in Fig. 1. In this scheme, the longest chain saturated fatty acid capable of forming oleic acid is decanoic acid and the longest chain saturated precursor of cis-vaccenic acid (Δ^{11}-C_{18}) is octanoic acid. This maximum chain length of precursors and lack of a requirement for oxygen clearly differentiates between this system and the one allowing direct dehydrogenation of a C_{16} or C_{18} saturated acid.

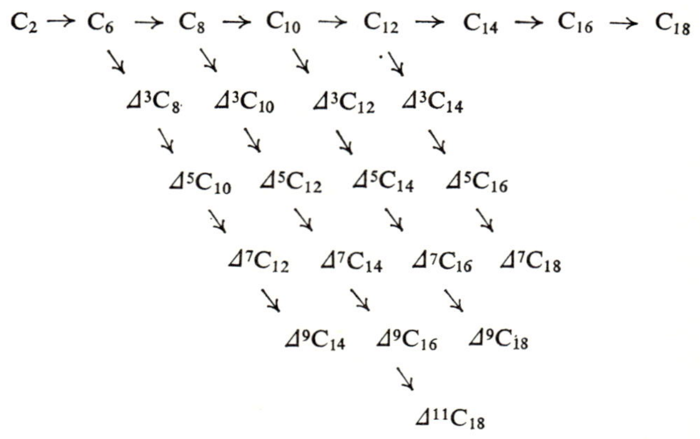

$$C_2 \rightarrow C_6 \rightarrow C_8 \rightarrow C_{10} \rightarrow C_{12} \rightarrow C_{14} \rightarrow C_{16} \rightarrow C_{18}$$
$$\searrow \quad \searrow \quad \searrow \quad \searrow$$
$$\Delta^3C_8 \quad \Delta^3C_{10} \quad \Delta^3C_{12} \quad \Delta^3C_{14}$$
$$\searrow \quad \searrow \quad \searrow \quad \searrow$$
$$\Delta^5C_{10} \quad \Delta^5C_{12} \quad \Delta^5C_{14} \quad \Delta^5C_{16}$$
$$\searrow \quad \searrow \quad \searrow \quad \searrow$$
$$\Delta^7C_{12} \quad \Delta^7C_{14} \quad \Delta^7C_{16} \quad \Delta^7C_{18}$$
$$\searrow \quad \searrow \quad \searrow$$
$$\Delta^9C_{14} \quad \Delta^9C_{16} \quad \Delta^9C_{18}$$
$$\searrow$$
$$\Delta^{11}C_{18}$$

FIG. 1.

The publication of Bloch's extensive studies on the biosynthesis of unsaturated acids, caused attention to be turned to plant systems. Mudd and Stumpf (1961) described the synthesis of oleic acid from acetate by the avocado mesocarp system and showed a clear requirement for molecular oxygen. James (1963), studying plant leaves, also established a requirement for oxygen in the synthesis of oleic and linoleic acids but direct dehydrogenation of palmitic and stearic acids could not be demonstrated. This same effect was shown to occur with isolated chloroplasts by Stumpf and James (1963). On the other hand, saturated acids of chain length up to C_{14} when supplied to leaf discs acted as direct precursors of the C_{18} unsaturated acids while chain lengths of up to C_{12} would act as precursors of the C_{16} unsaturated acids. This same effect was found by Stumpf (1967) in barley leaves. Since Bloomfield and Bloch (1960) had already shown that direct desaturation occurred in a blue-green alga, *Anabaena variabilis*, these organisms appeared to have a different pathway to that occurring in the leaves of higher plants.

The leaf pathway thus appeared as an exception and seemed to operate as follows:

$$C_2 \rightarrow C_8 \rightarrow C_{10} \rightarrow C_{12} \rightarrow C_{16} \text{ unsaturated acids}$$
$$\downarrow$$
$$C_{14} \rightarrow C_{18} \text{ unsaturated acids}$$
$$\downarrow$$
$$C_{16}$$
$$\downarrow$$
$$C_{18}$$

Similar results were obtained by Bloch and his coworkers (Erwin *et al.* 1964) with intact cells of the green alga, *Euglena gracilis*. However, James and coworkers (Harris *et al.* 1965) showed that whole cells of *Chlorella vulgaris* or *Euglena gracilis* were capable of directly desaturating palmitic and stearic acids to give the corresponding monoenoic acids when incubated in phosphate buffer. In a later paper, Nagai and Bloch (1965) showed that cell-free particulate extracts of etiolated *Euglena* were capable of directly desaturating the CoA thioesters of palmitic and stearic acids, but that soluble cell-free extracts from green cells grown in the light were capable of directly desaturating only the ACP thioesters. This suggested that the plant pathway also involved the direct desaturation of palmitic and stearic acids but that in some way added long chain acids were not being converted to the true enzyme substrates.

This interpretation was investigated by Harris *et al.* (1967), who incubated leaf discs with $2\text{-}^{14}C$-acetate under anaerobic conditions. After 6 h incubation, only palmitic and stearic acids were labelled because of inhibition of monoene synthesis by the lack of oxygen. Similar discs incubated anaerobically for 4 or 5 h were transferred after washing to aerobic conditions, and radioactive analysis of the labelled fatty acids gave the results shown in Table I. These results can be explained by a part of the anaerobically synthesized stearic acid being directly converted to oleic acid when oxygen was supplied. For this reason, the authors suggested that part of the stearic acid was bound to some acceptor that did not exchange with added stearate, and that this bound

TABLE I

Production of unsaturated long chain fatty acids from ^{14}C-acetate in leaves of *Spinacia oleracea* under aerobic and anaerobic conditions

			Conditions		
Distribution of activity in fatty acids		(1) 6 h anaerobic	(2) 5 h anaerobic + 1 h aerobic	(3) 4 h anaerobic + 2 h aerobic	(4) 6 h aerobic
Palmitic	% counts	50	53	48	35
	Specific activity	0·6	0·7	0·4	1·3
Stearic	% counts	50	34	35	4·7
	Specific activity	5·2	3·5	2·3	1·1
Oleic	% counts	0	13	17	52
	Specific activity	—	0·6	0·9	6·6
Linoleic	% counts	0	0	0	7·9
Specific activity		—	—	—	0·8

TABLE II

Desaturation of stearoyl-*S*-ACP by spinach chloroplasts in triethanolamine HCl buffer with 2-mercaptoethanol, spinach ferredoxin and NADPH (Nagai and Bloch, 1966)

Substrate	Percentage desaturation
Stearoyl-*S*-ACP	19·3
ditto	28·2
Stearoyl-*S*-ACP minus ferredoxin	2·6
Palmitoyl-*S*-ACP	1·8
Stearoyl-*S*-ACP	0·6

stearic acid was an effective substrate of a desaturase enzyme system essentially similar to that occurring in yeast, algae, etc. Subsequently, Nagai and Bloch (1966) showed that the enzyme substrate in both leaves of higher plants and green algae (*Euglena gracilis*) is the stearyl-*S*-ACP derivative (Table II). Thus the major difference between higher plants and algae must lie in the absence, in the former group, of the enzymes capable of converting the

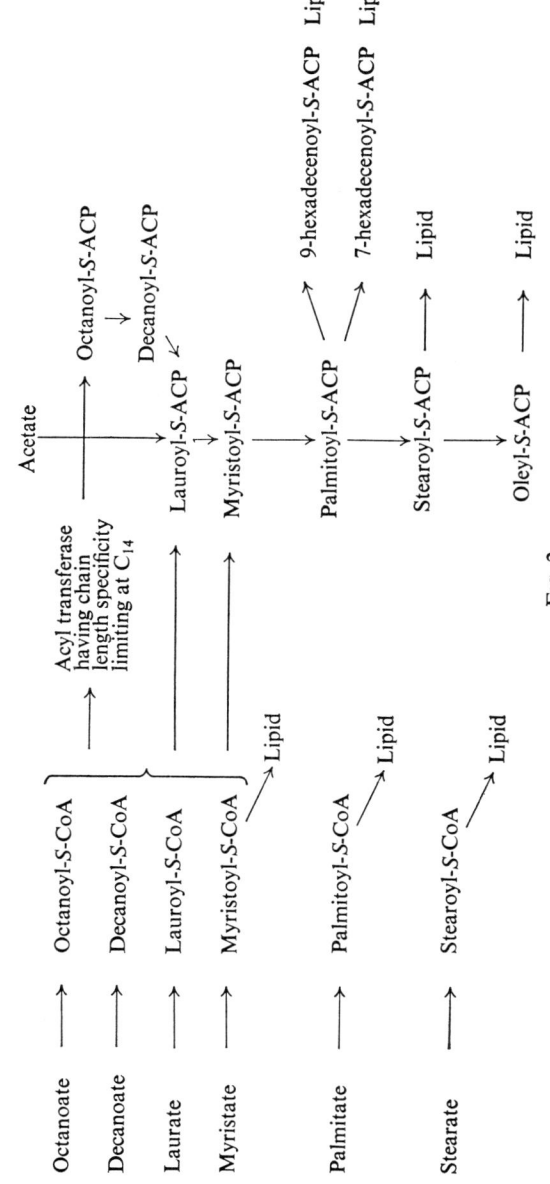

Fig. 2.

stearyl-S-CoA to stearyl-S-ACP. A schematic diagram covering these results with the leaf system is shown in Fig. 2.

B. CHAIN LENGTH REQUIREMENTS

Because the leaf system is unable to accept the free long chain acids as substrates, detailed work on the overall chemistry of desaturation has been done with the green algae.

Investigation of the direct dehydrogenation of a variety of chain lengths of saturated acids gave the results shown in Table III (Howling, Morris and James, 1968).

TABLE III

Conversion of C_{14}–C_{19} saturated acids to Monoenes by *Chlorella vulgaris*

Precursor	Monoenoic acids produced by *Chlorella vulgaris*	
14:0[a]	Δ_7, 14:1	Δ^9, 14:1
15:0	Δ_7, 15:1	Δ^9, 15:1
16:0	Δ_7, 16:1	Δ^9, 16:1
17:0	—	Δ^9, 17:1
18:0	—	Δ^9, 18:1
19:0	—	Δ^9, 19:1

[a] In this convention the first figure denotes the number of carbon atoms and the second the number of double bonds.

Fulco (1965) had already demonstrated that slices of the plant, *Carpobrotis chilense*, were able to convert myristic acid to the corresponding monoenoic acid as well as to longer chain saturated and unsaturated acids.

In *Chlorella vulgaris* there thus seem to be two desaturases, one specific for the 14, 15 and 16 carbon chains giving the Δ^7 acids and the other of much lower specificity capable of desaturating the 14, 15, 16, 17, 18 and 19 chains to give the 9-monoenoic acids. The Δ^7 desaturase gave much the same conversion for each chain length whereas the Δ^9 desaturase showed a marked peak with the 18 carbon acid, though this may have been largely due to the greater formation of the dienoic acid. Since the double bond position is clearly set with reference to the distance from the carboxyl group it is likely that this end of the chain is bound to the enzyme, presumably as a thiol ester.

C. STEREOCHEMISTRY

Schroepfer and Bloch (1965) showed with *Corynebacterium diphtheriae* that dehydrogenation of stearic acid to oleic acid involved the specific loss of the D-9 and D-10 hydrogen atoms. A similar approach adopted by Morris *et al.*

(1966) with the green algal system gave identical results in that the D-9 hydrogen atom was eliminated. In addition there was a *cis* elimination of the 9 and 10 hydrogen atoms, thus the D-10 hydrogen also was removed. The fatty acid chain must therefore be held in a specific orientation of the enzyme surface and since either the same or a similar enzyme is capable of desaturating the C_{14} chain it is unlikely that reactivity at the 9–10 position is induced by bending of the chain.

Bloch (1963) suggested that the fatty acid chain could form a loop so that the thiol ester group would be close to the C_9 and C_{10} hydrogen atoms and so participate in their labilization by a pseudo-transannular reaction. Elucidation of the detailed mechanism is unlikely until purer enzyme preparations are available that can be studied by modern physical techniques.

D. 3-*trans*-HEXADECENOIC ACID

This fatty acid originally described and defined by Debuch (1962), and Klenk and Knipprath (1962) was demonstrated by Haverkate and van Deenen (1965) to be unlike the other unsaturated acids in that it is esterified exclusively with phosphatidyl glycerol (II).

$$
\begin{array}{l}
CH_2 \cdot O \cdot CO \cdot R^1 \\
\quad | \\
R^2CO.O—CH \qquad O \quad CH_2OH \\
\quad\quad\quad | \qquad\quad | \quad | \\
\quad\quad CH_2 \cdot O \cdot —PO.CH \\
\quad\quad\quad\quad\quad\quad | \quad | \\
\quad\quad\quad\quad\quad\quad O^- \;\; CH_2OH
\end{array}
$$

(II) R^1 = linolenoyl, R^2 = 3-*trans*-hexadecenoyl

In addition they showed it to occur mainly in the 2-position in phosphatidyl glycerol from both leaves and green algae, the major species of the phosphatidyl glycerol of leaves being the 1-linolenoyl-2-(3-*trans*-hexadecenoyl)-phosphatidyl glycerol (II). The biosynthesis of this acid and its specific location therefore pose an interesting problem.

The precursor of 3-*trans*-hexadecenoic acid was shown by Nichols *et al.* (1965) to be palmitic acid, the reaction requiring not only oxygen but also light in both *Chlorella vulgaris* and in plant leaves. Incubation of generally labelled 3-*trans*-hexadecenoic acid with both systems showed it to be rapidly reduced to palmitic acid and the residual unreduced acid was then found to be distributed in all the acyl lipids present (Bartels, James and Nichols, 1967).

No specific acylating enzymes therefore exist to account for the highly specific location of this unsaturated acid. Inhibition studies with sterculic acid (III), itself a component of some seed oils, showed that the reaction palmitic acid → 7-*cis*-hexadecenoic acid + 9-*cis*-hexadecenoic was strongly inhibited but the reaction palmitic → 3-*trans*-hexadecenoic acid was unaffected (James

$$\text{CH}_3(\text{CH}_2)_7\text{—C}{=}\text{C—}(\text{CH}_2)_7\text{CO}_2\text{H} \quad \text{Sterculic acid} \quad \text{(III)}$$

et al. 1968). Furthermore sterculic acid did not inhibit the incorporation of palmitic acid into most of the lipids, so that its effect could not be exerted at the stage palmitic acid → palmitoyl-*S*-CoA. A logical explanation for these results would be:

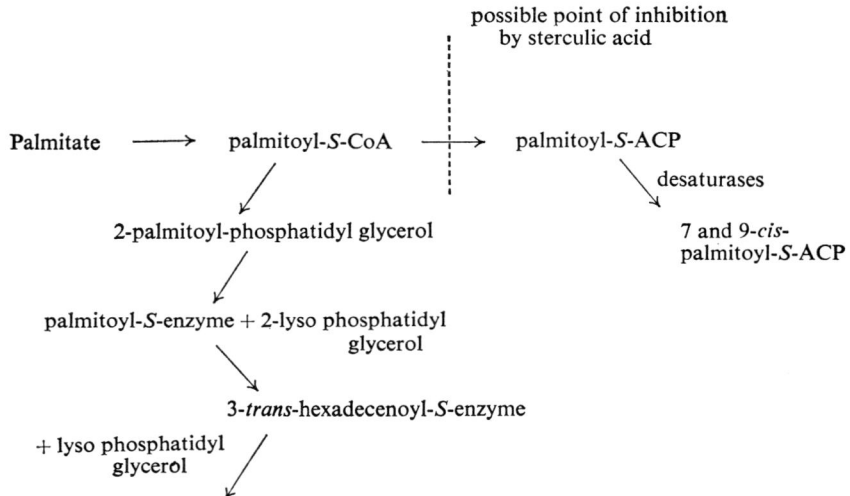

The possible involvement of acyl lipids as fatty acid carriers in other biosynthetic pathways will be discussed later.

E. POLYUNSATURATED ACIDS

Yuan and Bloch (1961) demonstrated that the yeast, *Torulopsis utilis* converted oleic acid to linoleic acid, the cofactor with a whole cell preparation being oxygen. Leaf discs also directly dehydrogenate oleic acid to give linoleic and linolenic acids (James, 1962) as do *Tetrahymena* species (Erwin and Bloch, 1963) and *Chlorella vulgaris* (Harris and James, 1965). The CoA ester of oleic acid is an effective substrate in cell free systems and oxygen and NADPH are required (e.g. in chloroplast preparations, Harris and James, 1965; and in safflower seed preparations, McMahon and Stumpf, 1964).

Investigations of the chain length specificity for conversion of monoenoic to dienoic acids gave the results shown in Table IV for *Chlorella vulgaris* (Howling, Morris and James, 1968).

Although both Δ^7 and Δ^9 monoenes are formed from myristic acid, no

TABLE IV

Conversion of saturated acids to dienoic acids by *Chlorella vulgaris*

Precursor	Double bond position in the dienoic acid produced of the same chain length	
14:0	0	0
15:0	0	9:12
16:0	7:10	9:12
17:0		9:12
18:0		9:12
19:0		9:12

dienoic acids are produced, the only Δ^7 acid giving rise to a 7:10 diene which has 16 carbon atoms. Whereas the enzyme or enzymes converting a Δ^9 acid to the 9:12 dienoic acid will accept 15, 16, 17, 18 and 19 C atom chains, there is a marked enhancement of conversion with the C_{18} chain.

F. STEREOCHEMISTRY

Morris *et al.* (1967) included in their stereochemical studies of dehydrogenation the conversion of monoenoic to di- and trienoic acids. The dehydrogenation of 12:13-dideutero-oleic to linoleic acid and 15:16-dideutero-oleic acid to linolenic acid by *Chlorella vulgaris* again demonstrated a *cis* elimination of the hydrogens from the 12:13 and 15:16 positions. Use of the D- and L-enantiomers of 12-tritio-stearic acid showed that the D-hydrogen atoms were removed at the 12 and 13 positions. By analogy the D-hydrogens from the 15 and 16 positions would also be removed.

G. POSSIBLE COUPLING OF FATTY ACID AND ACYL LIPID SYNTHESIS

Effective substrates for the desaturase systems so far demonstrated are the acyl-*S*-CoA thiol esters with etiolated *Euglena* cell-free systems (Nagai and Bloch, 1965) and the acyl-*S*-ACP thiol esters with green *Euglena* cell free preparations and spinach chloroplast preparations (Nagai and Bloch, 1966). Whether or not the acyl group is transferred from the CoA or ACP thiol esters to the desaturase enzymes is not yet known. If this is assumed then there is the possibility of other bound forms of fatty acid acting as acyl donor to the enzyme.

A detailed time study of the labelling of each of the component fatty acids synthesized from labelled acetate found in each of the eight lipids of *Chlorella vulgaris* was reported by Nichols *et al.* (1967). The bound fatty acids had a labelling-time relationship more reminiscent of active pools than of end products of lipid synthesis. The results suggested that initial esterification of certain intermediate fatty acids (myristic, palmitic, stearic and oleic acids) by specific lipids was followed by their removal and dehydrogenation and then by re-insertion into lipids of the same group. There was thus some similarity between the biosynthesis of the "conventional" unsaturated acids of the C_{16} and C_{18} series and that of 3-*trans*-hexadecenoic acid where an acyl lipid carrier seems to be involved.

However, it is possible that the acyl transfer system to form certain lipids is located spacially close to the synthetase and desaturase systems so that the transfers become specific to certain lipids. Definitive information on the coupling of fatty acid and acyl lipid synthesis will be obtained only by studying the utilization of acyl lipids by cell-free fatty acid synthesizing systems.

A summary of the present position, Fig. 3 gives the overall chemistry of synthesis of plant long chain fatty acids.

III. Unusual Fatty Acids

Seed oils are happy hunting grounds for those interested in the elucidation of structures of unusual fatty acids. In the past, ideas as to their biosynthesis have been largely limited to paper chemistry. Some biosynthetic work has now been carried out so that more definitive information is available on pathways.

A. RICINOLEIC ACID, D-12-HYDROXY-9-OCTADECENOIC ACID

The requirement for O_2 and NADPH in fatty acid desaturation has suggested to many workers, beginning with Bloch, that hydroxy acids were possible intermediates. As yet everyone has failed to demonstrate the conversion of a hydroxy acid to an unsaturated acid. However, ricinoleic acid, like linoleic acid, is formed in the seed of *Ricinus communis* from oleic acid (James, 1962; James, Hadaway and Webb, 1965; and Canvin, 1965). Further study with cell-free preparations from the same seed showed that oleyl-*S*-CoA was a suitable substrate and that molecular oxygen and NADPH were required (Yamada and Stumpf, 1964). No conversion of linoleic acid to ricinoleic acid could be demonstrated with either the intact or cell free systems. However in *Claviceps purpurea*, linoleic acid itself is the precursor of ricinoleic acid, and oxygen is not necessary as cofactor (Morris *et al.* 1966). The stereochemistry of the form-ation of ricinoleic acid has now been worked out by Galliard and Stumpf (1966) and by Morris (1967). Use of racemic *erythro*-12,13-ditritio-oleic acid with the intact seed system by Morris showed that only one H atom from the 12-position was removed during the conversion of oleic to ricinoleic acid. Furthermore use of D- and L-12-tritio-oleic acid demonstrated that the lost H atom had the D configuration. This showed that ricinoleic acid is formed by hydroxyl substitution at the 12-position with retention of configuration.

B. ACETYLENIC FATTY ACIDS

Bu'Lock and his group established that the carbon chains of several acetyl-enic compounds found in fungi were derived from acetate and malonate

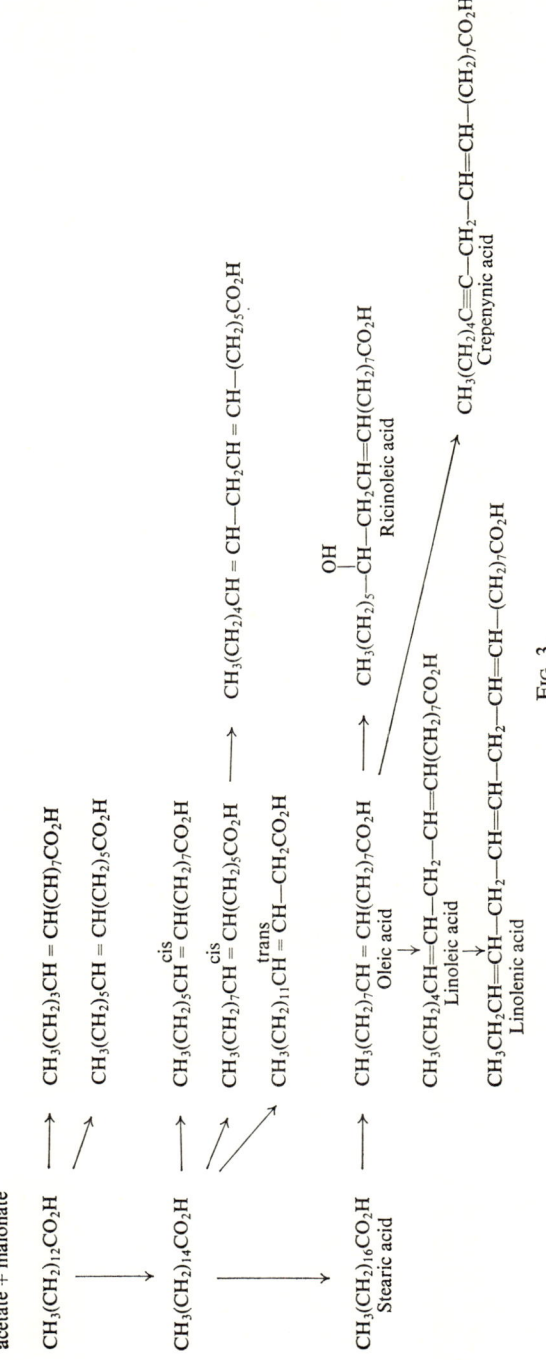

Fig. 3.

(Bu'Lock and Smalley, 1962). Similarly, ximenynic acid (*trans*-11-octadecen-9-ynoic) and related acids were formed from acetate by the plant *Santalum acuminatum* (Bu'Lock and Smith, 1963). Further papers described the conversion of ^{14}C-oleic acid to ^{14}C-crepenynic acid (*cis*-9-octadecen-12-ynoic, Fig. 3) by the fungus *Tricholoma grammopodium* (Bu'Lock and Smith, 1966) and Bu'Lock and Smith (1967). These authors believed that linoleic acid was an intermediate. Haigh *et al.* (1967) showed that seed of *Crepis rubra* converted 1-^{14}C-oleic acid to 1-^{14}C-crepenynic acid, molecular oxygen being required. Under identical conditions no conversion of linoleic acid to crepenynic acid could be demonstrated. Subsequent work (Haigh *et al.* 1968) demonstrated that neither *cis*, *trans* (*trans-cis*)-linoleic acid nor *cis*-12,13-epoxyoleic acid were precursors of crepenynic acid. On this evidence, introduction of the acetylenic bond at the 12–13 position appears to be a concerted reaction not involving prior formation of a double bond.

Oleic acid thus occupies a key position in fatty acid biosynthesis giving rise not only to the polyenoic acids but also to a variety of other acids of 18C chain length that have a double bond in the 9:10 position (Fig. 3).

IV. The Effect of Temperature on Fatty Acid Biosynthesis

Although it has been well documented over many years that a given species of plant will contain relatively more unsaturated fatty acids if grown at a lower temperature, no real explanation has been produced. Meyer and Bloch (1963) for example have shown that cell free systems obtained from *Torulopsis utilis* grown at 19° more rapidly desaturated stearoyl-*S*-CoA than did preparations from cells grown at 30°. Canvin (1963) studied the effect of growing temperature on the fatty acid composition of seeds of rape, safflower, sunflower, flax and castor. No effect was shown with the safflower and castor plants. The other species, however, did show a decrease in polyunsaturated acids and an increase in oleic acid at the higher temperature.

There are possibly two effects involved; (a) an increase in level of the desaturase enzymes after growth at a lower temperature and (b) a direct effect of the lower temperature on desaturation reaction velocity.

Tracer experiments to try to determine the existence of effect (b) have been carried out with photosynthetic (leaves and a green alga) and non-photosynthetic systems (storage tissue of bulbous plants) (James and Harris, 1968). It was found that the photosynthetic tissues did not respond to lower temperatures by increasing the rate of desaturation whereas the non-photosynthetic system did do so. The photosynthetic systems rapidly produce oxygen in the light, so oxygen is always available. Bulb tissue on the other hand has no endogenous oxygen supply and responded to an artificially increased level of oxygen in the gas phase at the highest temperature (40°) by an effect similar to that of low temperature (10°) alone. (The solubility of oxygen in water doubles from 40° to 10°.) The suggestion is therefore made that in some plant tissues oxygen is the rate determining factor in desaturation and that the fall in reaction velocity with drop in temperature is more than compensated for by the greater availability of oxygen. Indeed over extended periods this could possibly also affect the enzyme levels as well.

REFERENCES

Alberts, A. W., Goldman, P., and Vagelos, P. R. (1963). *J. biol. Chem.* **238**, 557.
Alberts, A. W., Majerus, P. W., Talamo, B., and Vagelos, P. R. (1964). *Biochemistry*, **3**, 1563.
Barron, E. J., Squires, C., and Stumpf, P. K. (1961). *J. biol. Chem.* **236**, 2610.
Bartels, C. T., James, A. T., and Nichols, B. W. (1967). *Europ. J. Biochem.* **3**, 7.
Bloch, K. (1963). *In* "The Control of Lipid Metabolism" (J. K. Grant, ed.), p. 1, Academic Press, London.
Bloomfield, D. K., and Bloch, K. (1960). *J. biol. Chem.* **235**, 337.
Brooks, J. L., and Stumpf, P. K. (1965). *Biochim. Biophys. Acta*, **91**, 213.
Brooks, J. L., and Stumpf, P. K. (1966). *Archs Biochem. Biophys.* **116**, 108.
Bu'Lock, J. D., and Smalley, H. M. (1962). *J. chem. Soc.* **4662**.
Bu'Lock, J. D., and Smith, G. N. (1963). *Phytochemistry*, **2**, 289.
Bu'Lock, J. D., and Smith, G. N. (1966). *Biochem. J.* **98**, 6P.
Bu'Lock, J. D., and Smith, G. N. (1967). *J. chem. Soc. (C)*, 332.
Canvin, D. T. (1963). *Can. J. Bot.*, **43**, 62.
Canvin, D. T. (1965). *Can. J. Bot.* **43**, 63.
Debuch, H. (1962). *Experentia*, **18**, 61.
Eberhardt, F. M., and Kates, M. (1957). *Can. J. Bot.* **35**, 907.
Erwin, J., and Bloch, K. (1963). *J. Biol. Chem.* **238**, 1618.
Erwin, J., Hulanicka, P., and Bloch, K. (1964). *Comp. Biochem. Physiol.* **12**, 191.
Fulco, A. J. (1965). *Biochim. Biophys. Acta*, **106**, 211.
Galliard, T., and Stumpf, P. K. (1966). *J. biol. Chem.* **241**, 5806.
Goldfine, H., and Bloch, K. (1961). *J. biol. Chem.* **236**, 2596.
Goldman, P., Alberts, A. W., and Vagelos, P. R. (1963a). *J. biol. Chem.* **238**, 1255.
Goldman, P., Alberts, A. W., and Vagelos, P. R. (1963b). *J. biol. Chem.* **238**, 8579.
Goldman, P., and Vagelos, P. R. (1962). *Biochem. Biophys. Res. Comm.* **7**, 414.
Haigh, W. G., Morris, L. J., and James, A. T. (1967). *Biochim. Biophys. Acta*, **137**, 391.
Haigh, W. G., Morris, L. J., and James, A. T. (1968) Lipids. (In Press).
Harris, R. V., Harris, P., and James, A. T. (1965). *Biochim. Biophys. Acta*, **106** 465.
Harris, R. V., and James, A. T. (1965). *Biochim. Biophys. Acta*, **106**, 456.
Harris, R.V., James, A. T., and Harris, P. (1967). *In* "Biochemistry of Chloroplasts", Vol. II, T. W. Goodwin, ed. Academic Press, London.
Haverkate, F., and van Deenen, L. (1965). *Biochim. Biophys. Acta*, **106**, 78.
Howling, D., Morris, L. J., and James, A. T. (1968). *Biochim. Biophys. Acta*, **152**, 224.
James, A. T. (1962). *Biochim. Biophys. Acta*, **57**, 167.
James, A. T. (1963). *Biochim. Biophys. Acta*, **70**, 9.
James, A. T. (1962). *Bull. Soc. Chim. Biol. Paris*, **44**, 951.
James, A. T., Hadaway, M. C., and Webb, J. P. (1965). *Biochem. J.* **95**, 448.
James, A. T., and Harris, P. (1968). *Biochem. J.* (In press).
James, A. T., Harris, P., and Bezard, J. (1968). *Europ. J. Biochem.* **3**, 318.
Klenk E., and Knipprath, W. (1962). *Z. Phys. Chem.* **327**, 283.
Lennarz, W. J., Light, R.J., and Bloch, K. (1962). *Proc. Natl. Acad. Sci. U.S.* **48**, 840.
McMahon, V., and Stumpf, P. K. (1964). *Biochim. Biophys. Acta*, **84**, 359.

Meyer, F., and Bloch, K. (1963). *Biochim. Biophys. Acta*, **77**, 671.

Morris, L. J. (1967). *Biochem. Biophys. Res. Comm.* **29**, 311.

Morris, L. J., Hall, S. W., and James, A. T. (1966). *Biochem. J.* **100**, 29C.

Mudd, J. B., and McManus, T. T. (1962). *J. biol. Chem.* **237**, 2057.

Mudd, J. B., and Stumpf, P. K. (1961). *J. biol. Chem.* **236**, 2602.

Nagai, J., and Bloch, K. (1965). *J. biol. Chem.* **240**, 3702.

Nagai, J., and Bloch, K. (1966). *J. biol. Chem.* **241**, 1925.

Nichols, B. W. (1965). *Biochem. Biophys. Acta*, **106**, 274.

Nichols, B. W., Harris, P., and James, A. T. (1965). *Biochim. Biophys. Res. Comm.* **21**, 473.

Nichols, B. W., James, A. T., and Breuer, J. (1967). *Biochem. J.* **104**, 486.

Overath, P., and Stumpf, P. K. (1964). *J. biol. chem.* **239**, 4103.

Scheuerbrandt, G., and Bloch, K. (1962). *J. biol. Chem.* **237**, 2064.

Schroepfer, G. T., and Bloch, K. (1965). *J. biol. Chem.* **240**, 54.

Smirnov, B. P. (1960). *Biokhimija*, **25**, 419.

Smirnov, B. P. (1962). *Biokhimija*, **27**, 127.

Smirnov, B. P., and Rodionov, M. A. (1964). *Biokhimika*, **29**, 335.

Stumpf, P. K. (1967). *In* "Biochemistry of Chloroplasts", Vol. II (T. W. Goodwin, ed.) p. 213. Academic Press, London.

Stumpf, P. K., Bove, J. M., and Goffeau, A. (1963). *Biochim. Biophys. Acta*, **70**, 260.

Stumpf, P. K., and James, A. T. (1963). *Biochim. Biophys. Acta*, **70**, 20.

Vagelos, P. R., Majerus, P. W., Alberts, A. W., Larrabee, A. R., and Ailhaud, G. P. (1966). *Fed. Proc.* **25**, 1485.

Yuan, C., and Bloch, K. (1961). *J. biol. Chem.* **236**, 1277.

Yamada, M., and Stumpf, P. K. (1964). *Biochem. Biophys. Res. Comm.* **14**, 165.

CHAPTER 5

Recent Developments in Molecular Taxonomy

H. ERDTMAN

Department of Organic Chemistry, Royal Institute of Technology, Stockholm
Sweden

I.	Introduction .	. 107
II.	Fossils Ancient and "Modern"	. 108
III.	Comparative Phyto- and Zoochemistry	. 109
	A. Compilations	. 109
	B. Screenings of Taxa	. 109
IV.	Biosynthesis and Molecular Taxonomy	. 112
V.	Molecular Biology and Molecular Taxonomy	. 118
VI.	The Future	. 118
	References	. 119

I. Introduction

Ten years ago the term Chemotaxonomy was seldom found in the organic chemical literature concerned with natural products. In those days the isolation and the structural and configurational elucidation of compounds present in natural materials were the main occupation of most organic chemists with an interest in this branch of organic chemistry. The objects of their researches were usually chosen more or less by chance as providing purely chemical problems without reference to their biological significance.

Until recently, the structural elucidation of natural products was a costly and time-consuming affair. This situation has now changed radically with the introduction of many powerful physical methods, and our knowledge of the chemistry of plants and animals is presently increasing at a fantastic rate. As a result there is a rapidly changing outlook in natural product chemistry. The elucidation of the structure of a natural product is no longer regarded as an end in itself but as a means of clarifying a biological problem. The study of the distribution of natural products for taxonomic purposes is an example, and chemotaxonomy is presently attracting much interest not only among the biologists, but also among chemists.

As indicated by the word itself, chemotaxonomy is a hybrid science. There has been some dispute as regards the suitability of the term. Some prefer "chemical systematics" but this is scarcely acceptable to chemists who have their own systematic problems. Others recommend "biochemical systematics". The reason for these differences in opinion seems to be that in some circles the term "taxonomy" is used, somewhat incorrectly, as including "systematics" whilst in others the two terms are distinctly separated. Personally, I prefer "molecular taxonomy" because what we are dealing with is very definitely the distribution of compounds having a discrete molecular structure in relation to taxonomy (or systematics).

II. Fossils Ancient and "Modern"

Unfortunately, fossils do not normally lend themselves to chemical investigations aimed at discovering constituents of taxonomic interest. However, many organic compounds, for example amino acids, have been isolated from fossils and biogenic deposits obtained from different geological periods, the oldest dating back some 2000–3000 million years. Organic geochemistry is now expanding rapidly and in future the results may become taxonomically valuable. Abelson (1959) has summarized what was known up to 1959 and a more detailed discussion Eglinton and Murphy, is expected to appear shortly (see also Horowitz and Miller, 1962).

Some information on the chemistry of organisms of past geological periods may be inferred from investigations of modern plants and animals belonging to recognized ancient groups. *Psilotum nudum* and *Tmesipteris tannensis* belonging to the two recent genera of the Silurean-Devonian Psilophytales both contain psilotin (I), a dihydropyrone glucoside (Tse and Towers, 1967). This compound is structurally related to 6-phenylcoumalin and paracotoin from

Psilotin (I)

Aniba (*Nectandra*), Lauraceae. *Ginkgo biloba* is another "living fossil" producing among other interesting secondary products the recently elucidated ginkgolides (Woods *et al.* 1967). These are certainly remarkable compounds, but not more so than many other constituents of plants which are less ancient. The coelacanth *Latimeria* is famous because it was formerly believed to be extinct. The primary metabolites of *Latimeria* are, as might have been expected, similar to those of modern fish but Haslewood (1967) has shown that its bile acids are "primitive", i.e. closer to cholesterol than those of higher vertebrates.

III. COMPARATIVE PHYTO- AND ZOOCHEMISTRY

A. COMPILATIONS

Our accumulated knowledge of the chemistry of plants and animals forms the basis of molecular taxonomy. It is thus very desirable for biologists, as well as chemists, to be able to orient themselves rapidly among the chemical constituents which have been found either in a certain taxon or in related taxa. For this reason, compilations of natural products, listed according to variation in chemical structure and giving the sources from which the compounds have been obtained, are indispensable. They can be even more useful if the information is organized according to some biological system. The latter has been done in Hegnauer's monumental "Chemotaxonomie der Pflanzen" of which the first volume appeared in 1962. All those interested in the chemical approach to plant systematics are grateful to Professor Hegnauer for collecting, organizing and discussing all the essentials of the phytochemical literature and pointing out the greatest gaps in our knowledge. Whilst Hegnauer's work almost amounts to a phytochemical Beilstein and hopefully will be followed by "Ergänzungsbände", Kariyone's "Annual Reports on Plant Chemistry" is a sort of phytochemical "Chemical Abstracts". Mentzer and Fatianoff (1964) made a valiant attempt to organize the recent literature in this field according to "accepted" biogenetic hypotheses in their "Actualités de Phytochimie fondamentale", of which two further volumes have now appeared (Mentzer, 1966, 1968).

Another work which has been of value for the recent development of molecular taxonomy is the "Chemical Plant Taxonomy" edited by Swain (1963). As the Editor says in the preface this is "the first comprehensive attempt to survey the scope and usefulness of chemical plant taxonomy". Its form was determined by a symposium held in Paris in 1962. That symposium was important because it brought together scientists of different specializations, botanical taxonomists, phytochemists and chemists interested in biosynthetic problems. The volume is rather heterogeneous but it laid a foundation on which to begin to build bridges between the disciplines. "Comparative Phytochemistry" also edited by Swain (1966), "Biochemistry of Phenolic Compounds" edited by Harborne (1964) and Alston and Turner's "Biochemical Systematics" (1963) follow similar lines.

B. SCREENINGS OF TAXA

Systematic investigation of the distribution of chemical compounds in related taxa had been started by some organic chemists much earlier. Our own work in Stockholm on the conifers originated from the known presence of pinoresinol in spruce as well as in pine, and from considerations of the biosynthesis of lignans and lignins.

Sörensen, in Trondheim, began his pioneering studies on the polyacetylenes of Compositae about 1940 with an investigation of the "matricaria ester". His work followed systematic lines from the beginning and led to the discovery of a large number of polyacetylenes in different tribes of this family (Sörensen, 1963). Later, Bohlmann (1967) entered the same field and we owe to him and his coworkers much of our present knowledge about the distribution of the acetylenes in the Compositae, and about the many heterocyclic compounds belonging to the same series. Polyacetylenes were also found in *Cicuta virosa* and *Oenanthe crocata* of the Umbelliferae, and Bohlmann has extended his systematic investigations to include that family and the related Araliaceae with great success.

The occurrence of polyacetylenes in Compositae and Umbelliferae is interesting because some systematists have considered the phylogenetic relationship between these families. However, as Sörensen (1963) puts it:

> ... "in a field which is showing such a tremendous and rapid progress (as chemotaxonomy) there is every reason to restrict conclusions to the smaller taxonomic problems ...".

This is a word of warning which is augmented by the more recent discovery of polyacetylenes in Leguminosae and in Pittosporaceae (see Chapter 6). Simple acetylenic C_{18}-acids have long been known to occur, for example in Santalaceae and Oleaceae, and it would appear that they are fairly widely distributed, albeit easily overlooked.

The fact that polyacetylenes and polyacetylene derivatives such as mycomycin, terthienyl and junipal were found to occur in various fungi caused surprise because there were obviously no taxonomic relations between these organisms and the above mentioned taxa. Sir Ewart Jones (1960, 1966) and his school have made extensive investigations on these fungal polyacetylenes.

About 1950, Kjaer commenced a systematic screening of Cruciferae as well as the related families, Capparidaceae, Resedaceae, Moringaceae and Tovariaceae for mustard oils producing glucosinolates. Apparently all species of these related families produce glucosinolates, compounds of the general structure

$$\text{Glc-S}-\overset{\displaystyle R}{\underset{|}{C}}=\text{N}-\text{O}-\text{SO}_3{}^-$$

(II)

(II) in which R can vary considerably. However, compounds of this type have never been found in any of the Papaveraceae. This is of interest since some botanists place Papaveraceae close to Cruciferae whilst others believe that Papaveraceae has an entirely different ancestry (Kjaer, 1963; Ettlinger and Kjaer, 1968). As in so many other cases, some quite unrelated genera also contain glucosinolates, *Tropaeolum* being a well known example.

The heart-active cardenolide glycosides also occur in unrelated plants, but they are particularly common in the closely related families Apocynaceae and

Asclepiadaceae. The breath-taking, systematic work of Reichstein, especially his screening of the genus *Strophanthus* (Apocynaceae, subfamily Apocynoideae) for cardenolide glycosides constitutes an outstanding example of systematic plant chemical investigations (see e.g., Geiger *et al.* 1967).

The pioneering studies of Kjaer, Reichstein and Sörensen, are typical of ambitious, recent phytochemical investigations with a taxonomic slant. These chemists have all concentrated upon groups of secondary plant constituents of general occurrence in the relevant genera and all of them are almost professional taxonomists.

The Apocynaceae, particularly the subfamily Plumerioideae, are rich in alkaloids, mostly complex indole derivatives. Djerassi has, for a long time, studied the chemistry of the *Aspidosperma* alkaloids and their distribution in this apocynaceous genus (see Arndt *et al.* 1967). In one of the early systematic investigations of this kind Djerassi (1957) also studied the triterpenes of the Cactaceae, plants known to produce several characteristic alkaloids.

As has often been pointed out it is, however, also important to investigate as many compounds of as different biosynthetic origin as possible in each taxon. Patterns of constituents are taxonomically more useful than single chemical characteristics. It is for that reason that we have looked, not only for simple phenols but also for flavonoids, tropolones and terpenoids in our conifers.

In Engler's "Syllabus der Pflanzenfamilien" (1964) the Cactales are placed close to Centrospermae and this relationship is amply confirmed by the occurrence of "betalains" in both orders. Dreiding's structural elucidation of betanin (III), a betalain from red beet and the extensive studies of the distribution of these compounds are not only brilliant chemical achievements but also

Betanin (III)

taxonomically very important (Wilcox *et al.* 1965; Mabry, 1966). Apparently betalains occur in all Centrospermae with the exclusion of the Caryophyllineae. The latter taxon has been considered to be the most highly developed of the Centrospermae. It would appear strange if just the Caryophyllineae had switched over from the production of betalains to that of the common angiosperm anthocyanins. We have here a conflict between chemical and classical taxonomy which should stimulate further inquiries into the chemistry and systematics of the relevant taxa. It is an intriguing fact that the betalains are produced to the exclusion of anthocyanins. This constitutes an interesting biochemical problem.

Many other recent studies on the chemistry of genera or higher categories could be mentioned but this would take too long. Let me only mention Bate-Smith's extensive studies on the distribution of various plant phenolics; Harborne's, Hörhammer's and Wagner's work on flavone glycosides, normal O- as well as C-glycosides; Ourisson's investigations of the genus *Hypericum* and the family Dipterocarpaceae; Herout's and Sorm's researches on sesquiterpenes from *Petasites* and other Compositae (see Chapter 7), to which Herz has added much information; and Price's extensive studies of the rutaceous, and Lederer's on the simaroubaceous, alkaloids.

It is not too great an exaggeration to say that simply by looking at the structure of a new natural product, one is now able to make a fair guess as to the genus or family from which the compound was obtained. At least I often indulge in this form of entertainment.

As bits of information are slowly being put together by different research workers, the "chemical profiles" of even large plant families such as Compositae and Leguminosae are becoming increasingly apparent. Orchidaceae, perhaps the largest of them all, was long supposed to be chemically rather uninteresting. However, these plants have recently been found to produce many alkaloids and they are now being systematically and actively investigated, among others, by Lüning in Stockholm. Judging by what has already emerged, Orchidaceae will certainly afford plenty of surprises.

Taxonomists recognize evolutionary progressions and regressions and it is obviously important for chemists to look for possible relationships between such evolutionary trends and chemistry. Goldschmidt has recently discussed the chemistry of the genus *Aniba*, Lauraceae, from this point of view (see Mors *et al.* 1962).

Secondary metabolites have proved to be highly useful as chemical characteristics of plants; much less so of animals. However, recent studies on the bile salts of vertebrates (Haslewood, 1967), the branched fatty acids of the preen glands of birds (Odham, 1966), the toxic principles of arthropods (Weatherston, 1967) and insect pheromones possess considerable interest in relation to the taxonomy of these groups and activity in these fields is at present increasing rapidly.

IV. BIOSYNTHESIS AND MOLECULAR TAXONOMY

It has long been clear that the comparison of chemical characteristics can lead to biologically absurd conclusions, particularly if single constituents are used. One of the main reasons for such calamities is the failure to realize that chemically identical compounds may be "biologically different". Chemically identical compounds are sometimes, probably frequently, synthesized in different ways in unrelated organisms. In other words, the complex biosynthetic processes have a higher taxonomic value than the secondary products themselves.

There are also several indications that in any given species there may exist several pathways leading from simple ubiquitous metabolites, as well as from more advanced intermediates, to the same accumulated "final" product. As a matter of fact it is just as difficult to define a chemical compound in a biologically exact way, as it is to define a species. Nevertheless both concepts are biologically useful.

To the many difficulties inherent in molecular taxonomy one could add that we know almost nothing about the chemical activities of the individual cells of a seemingly homogeneous tissue except, perhaps, for some highly specialized organs such as the terpene producing glands of, for example, *Mentha*. As a rule several chemically related compounds co-occur in the same organs such as wood or bark. Are they formed in the same or in different cells?

Finally, chemists must not forget that, as the result of evolutionary convergence, identical or related compounds may be produced in more or less similar ways in unrelated taxa.

It is obvious that comparative phytochemistry needs to be complemented by studies on the biosynthesis of secondary metabolites. A large volume of biosynthetic investigations has recently been carried out aiming at testing experimentally the numerous biogenetic or "intuitive" hypotheses which have been advanced on the grounds of comparisons of molecular architecture and known biochemical reactions of great generality. This has put our knowledge of biosynthetic routes on a much firmer basis than before.

Biosynthetic studies using labelled compounds, mutants or occasionally specific antagonists, e.g. ethionine as an antagonist of methionine, have a great interest in themselves but they have also given results of considerable taxonomic value. The exemplary and meticulous work of Barton, Battersby and their collaborators on the biosynthesis of morphine can be mentioned as an example (Barton *et al.* 1965). Tyrosine is transformed into (−)-reticuline (IV) which is then oxidized to (+)-salutaridine (V). Reduction and elimination of water follows giving (VI) leading to thebaine (VII) which is demethylated and via codeinone (VIII) reduced to codeine (IX) and demethylated to morphine (X).

The transformation of tyrosine to benzylisoquinoline seems to be a relatively simple process requiring enzymes common to many unrelated plants and, for example, reticuline was isolated from *Annona reticulata*, Annonaceae, before it was found to occur in the opium poppy. Oxidative phenol couplings, inter- as well as intra-molecular, are also common biosynthetic reactions. Salutaridine is produced not only by *Papaver somniferum* but also by *Croton salutaris*, Euphorbiaceae. The later steps in the synthesis are more species specific and the end-product, morphine, is a unique compound.

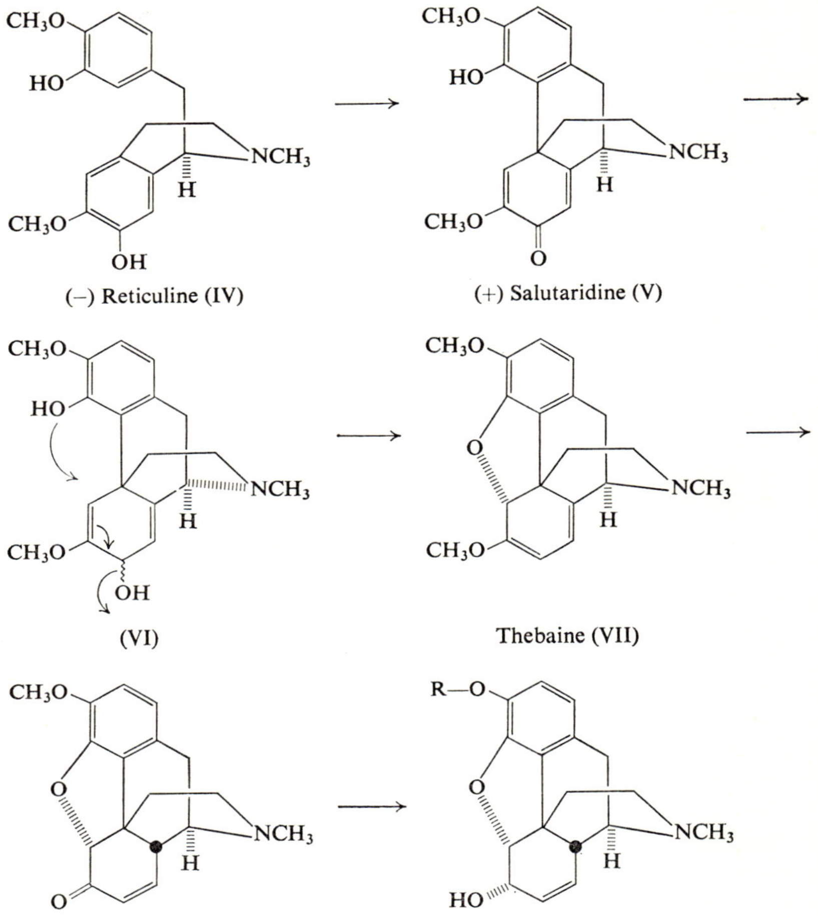

(−) Reticuline (IV)

(+) Salutaridine (V)

(VI)

Thebaine (VII)

Codeinone (VIII)

Codeine (R = CH$_3$) (IX)
Morphine (R = H) (X)

A dienone derived from the novel C$_{17}$-phenols occurring in several conifers is athrotaxin which has been isolated from the heartwood of *Athrotaxis selaginoides*, Taxodiaceae. Its structure (Nishimura, 1966) is almost certainly (XI) and the compound could be a dehydrogenation product of a dihydroxy-sugiresinol analogue (XII). Hydroxysugiresinol (XIII) is one of the C$_{17}$-phenols accompanying athrotaxin in the wood.

The beautiful and stringently executed researches of Battersby (1967) and his colleagues on the biosynthesis of colchicine (XV) from *Colchicum* are also of great taxonomic interest. It was a fair guess that at some stage in the biosynthesis a phenol coupling took place. The formation of the tropolone ring appeared to proceed, as is assumed for the conifer tropolones, via the in-

Athrotaxin (XI)

(XII)

Hydroxysugiresinol (XIII)

corporation of an adjoining carbon atom into an aromatic ring, but how and when was impossible to predict. The mist dispelled when a "missing link", androcymbine (XIV) from *Androcymbium melanthioides*, a close relative of *Colchicum*, was discovered. Androcymbine is a congener of colchicine in *Androcymbium* and is transformed in high yield into colchicine by *Colchicum autumnale*. Androcymbine is probably not entirely strange to this plant. Androcymbine and colchicine are very different from a structural point of view; nevertheless they are biologically closely related.

Androcymbine (XIV) → Steps → Colchicine (XV)

In androcymbine a $C_6C_3NC_2C_6$ unit plays an essential role while the Amaryllidaceae alkaloids possess a $C_6C_2NC_1C_6$ unit. In striking contrast are the very common $C_6C_2NC_2C_6$ alkaloids of the dicotyledons. The mesembrine alkaloids, e.g. mesembrine (XVI) from *Sceletium* (*Mesembryanthemum*) species, Aizoaceae (Centrospermae) are mostly regarded as being biosynthetically related to the Amaryllidaceae alkaloids (compare haemanthamine (XVII))

Mesembrine (XVI)   Haemanthamine (XVII)

but one ring carbon atom is missing, and Amaryllidaceae is a monocotyledonous family while Aizoaceae is a dicotyledonous one. These are not, however, decisive arguments against a biosynthetic relationship and further studies on the biosynthesis (Jeffs *et al.* 1967) of the mesembrine alkaloids are awaited with interest.

The biosynthesis of nicotine in *Nicotiana* (Solanaceae), has been much investigated. This relatively simple alkaloid is structurally similar to anabasine from *Anabasis aphylla*, Chenopodiaceae. The latter alkaloid is also produced by some *Nicotiana* species, e.g. *N. glauca*. The piperidine ring of anabasine from *N. glauca* is believed to originate from lysine and biosynthetic investigations lend support to this assumption. Similarly ornithine might be expected to give rise to the pyrrolidine ring in nicotine and there is also biosynthetic support for this. However Rapoport followed the path of carbon in nicotine synthesis and obtained results which differed considerably from those from feeding experiments (Liebman *et al.* 1967). It is possible that these conflicting results are due to the occurrence of several alternative pathways to the pyrrolidine-ring precursor or precursors. The nicotine problem is interesting because it emphasizes that the biosynthesis of natural products ought to be studied with different techniques. This may lead to discoveries which cannot easily be made using labelled hypothetical precursors.

Coniine (XVIII) from *Conium maculatum*, Umbelliferae, has been supposed to be derived from lysine, but Leete (1963, 1964) showed that it is of acetate (polyketide) origin. However, the chemically related *N*-methylisopelletierine (XIX) from *Sedum sarmentosum*, Crassulaceae, is derived from lysine, as is sedamine (XX) from *Sedum acre* (Gupta and Spencer, 1968). Hence, there are great biosynthetic differences between these simple, chemically similar alkaloids.

Coniine (XVIII)

Cholesterol is synthesized in animals from squalene-2,3-oxide via lanosterol, but in the biosynthesis of phytosterols cycloartenol apparently replaces lanosterol (Goad, 1967). Hence there should be characteristic biological differences between these steroids. However, cholesterol has been found to occur in

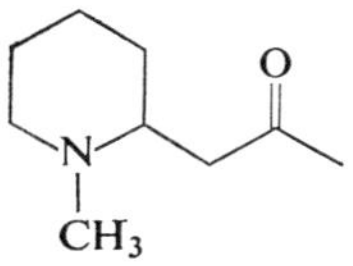

N-methylisopelletierine (XIX)

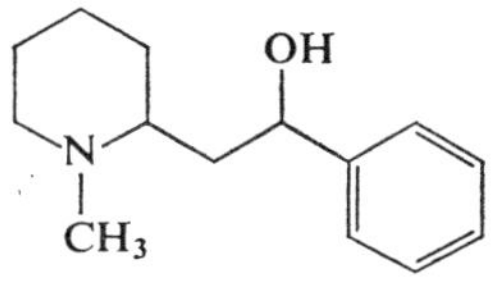

Sedamine (XX)

several plants and lanosterol in yeast. This is an exciting situation and is presently attracting much interest.

Whilst the fungal anthraquinones endocrocin and emodin are of polyketide origin, evidence is now accumulating which indicates that many anthraquinones from higher plants are formed in a different way and that the useful polyketide hypothesis must be employed with discrimination. This has long been suspected. Among other similar compounds teak wood, *Tectona grandis*, Verbenaceae, contains tectoquinone (XXI) and deoxylapachol (XXII) and, as

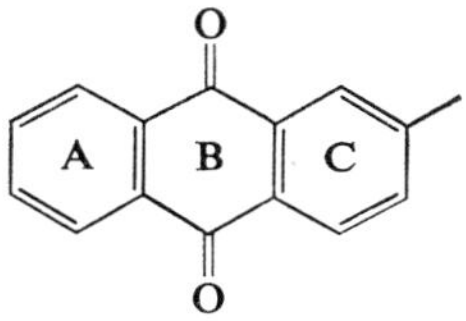

Tectoquinone (XXI)

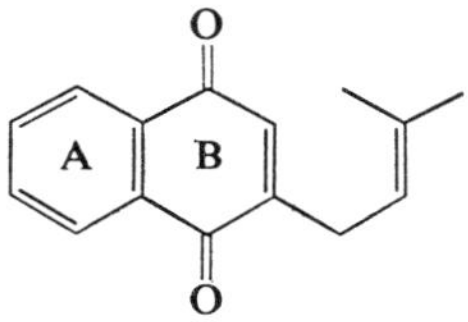

Deoxylapachol (XXII)

Sandermann and Simatupang (1966) have pointed out, this might indicate that ring C is derived from mevalonic acid. Using feeding techniques Leistner and Zenk (1968) have obtained results showing that ring A of alizarin (XXIII) from *Rubia tinctorum* (Rubiaceae) is built up from shikimic acid and that ring C is of acetate origin.

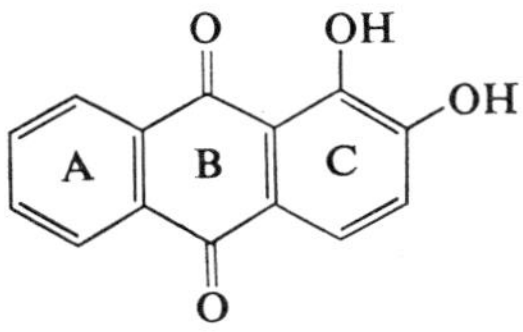

Alizarin (XXIII)

Important support for these ideas has been obtained by Burnett and Thomson (1968a) who isolated prenylated naphthalene derivatives as well as anthraquinones from the heartwood of *Tabebuia avellanae*, Bignoniaceae. Similar results have emerged from studies of several *Galium* species (Rubiaceae) and of *Pyrola*, Pyrolaceae (Burnett and Thomson, 1968b,c).

Murrayanine (XXVI) from *Murraya koenigii*, glycozoline and glycozolidine from *Glycosmis pentaphylla* and heptaphylline from *Clausena heptaphylla*,

5

three very closely related genera of the Rutaceae, could be heterocyclic analogues of tectoquinone and derived, as indicated schematically (XXIV, XXV), from 3-prenylated quinolines which are common in this family. Alternatively, a 2-prenylated indole could serve as an intermediate.

(XXIV)   (XXV)   Murrayanine (XXVI)

Xanthones are obviously also formed in different ways in different organisms. There is much biosynthetic evidence that shows that they are formed in fungi from one or more polyketide chains but, for example, in Guttiferae (*Kielmeyera*, *Mesua*), Moraceae (*Maclura*) and Gentianaceae they appear to be formed by intramolecular C–O–C-coupling of benzophenones, partly of acetic acid, partly of shikimic acid origin (Lewis and Warrington, 1964; Wolfrom and Bhat, 1965; Locksley *et al.* 1967).

V. MOLECULAR BIOLOGY AND MOLECULAR TAXONOMY

Molecular biologists have made spectacular advances in exploring the "machines and factories" of the cell community in which the common metabolites are manufactured, and the ways in which these subcellular units are organized, co-ordinated and genetically controlled.

The elucidation of the structures of certain proteins and polypeptides, e.g. cytochromes and haemoglobins, from different taxa has shown that they contain segments possessing identical amino acid sequences and that by comparing them there may be a possibility of tracing phylogenetic relationships. This has led to much speculation as to the possibilities of "reconstructing" the amino acid sequences in ancestral organisms and, ultimately, the evolutionary processes themselves as well as the time that it has taken to pass from primitive forms of life to the present ones. Not overburdened by classical evidence, some molecular biologists, and even scientists from quite different fields, have questioned some of the fundamental principles of the evolutionary theory (c.f. Moorhead and Kaplan, 1967). Many of the arguments are well known to biologists from discussions over seventy years ago.

Molecular biology tends to diverge considerably from other life-sciences and I feel that chemists and biologists should in the future forge closer connections with the strong and successful group of molecular biologists.

VI. THE FUTURE

The great potentialities of molecular taxonomy are now manifest to most biologists and chemists, as are the numerous pitfalls. The communication barriers between the disciplines are slowly being abraded and biologists are employing chemical methods with greater exactness and interpreting their results with greater care. Chemists are becoming increasingly aware of the necessity of exactly defining their biological material and are beginning to appreciate the nomenclatural difficulties with which biologists have to struggle. In their publications, the names of the taxonomic authorities are now more frequently added to the generic names and their meaning and importance is understood. Still it does happen that chemists work with identical plants described under different, synonymous, names trying to discover differences between them. More often than before, the family name is added to the name of the species. This considerably facilitates the chemist's appreciation of taxonomic relations.

It has been said that chemotaxonomy has come of age. This may be true. Its development during the last ten years or so has been truly remarkable. Phytochemistry, pure and comparative, has advanced rapidly and so has our knowledge of the biosynthesis of many types of natural products. However, considering the fact that only a small percentage of the extant organisms have yet been subjected to any chemical investigation at all, we have no reason to jubilate and exclaim "wie herrlich weit wir es gebracht". We are only at the very beginning.

The future of molecular taxonomy is bright. If it has not yet come of age, it may do so within another ten years.

REFERENCES

Abelson, P. H. (1959). *Fortschr. Chem. org. Naturstoffe*, **17**, 379.

Alston, R. E., and Turner, B. L. (1963). "Biochemical Systematics", Prentice Hall, New Jersey, U.S.A.

Arndt, R. R., Brown, S. H., Ling, N. C., Roller, P., Djerassi, C., Farreira, J. M., Gilberg, F. B., Miranda, E. C., Flores, S. E., Duarte, A. P., and Carrazsoni, E. P. (1967). *Phytochemistry*, **6**, 1653.

Barton, D. H. R., Kirby, G. W., Steglich, W., Thomas, G. M., Battersby, A. R., Dobson, T. A., and Ramuz, H. (1965). *J. chem. Soc.* 2423.

Battersby, A. R. (1967). *Pure appl. Chem.* **14**, 117.

Bohlmann, F. (1967). *Fortschr. Chem. org. Naturstoffe*, **25**, 1.

Burnett, A. R., and Thomson, R. H. (1968a). *J. chem. Soc.* (C) 850.

Burnett, A. R., and Thomson, R. H. (1968b). *J. chem. Soc.* (C) 854.

Burnett, A. R., and Thomson, R. H. (1968c). *J. chem. Soc.* (C) 857.

Djerassi, C. (1957). *In* "Festschrift A. Stoll," p. 330, Birkhäuser Verlag, Basel.

Ettlinger, M., and Kjaer, A. (1968). *In* "Recent Advances in Phytochemistry" (T. J. Mabry, R. E. Alston and V. C. Runeckles, eds.), p. 89, Appleton Century Crofts, New York, U.S.A.

Geiger, U. P., Weiss, S. K., and Reichstein, T. (1967). *Helv. chim. Acta*, **50**, 179; and numerous earlier papers.

Goad, L. J. (1967). *In* "Terpenoids in Plants" (J. B. Pridham, ed.), p. 159, Academic Press, London and New York.

Gupta, R. N., and Spencer, J. D. (1968). *Chem. Comm. London*, 85.

Harborne, J. B. (ed.) (1964). "Biochemistry of Phenolic Compounds", Academic Press, London and New York.

Haslewood, G. A. D. (1967). "Bile Salts", Methuen, London.

Hegnauer, R. (1962–1966). "Chemotaxonomie der Pflanzen", Bd. I–IV, Birkhäuser Verlag, Basel.

Horowitz, N. H., and Miller, S. L. (1962). *Fortschr. Chem. org. Naturstoffe*, **20**, 423.

Jeffs, P. W., Archie, W. C., and Farrier, D. S. (1967). *J. Am. chem. Soc.* **89**, 664.

Jones, Sir E. R. H. (1960). *Proc. Chem. Soc. (London)*, **1** 99.

Jones, Sir E. R. H. (1966). *Chem. Brit.* **2**, 6.

Kariyone, T. (ed.) (1964). Annual Index of the Reports on Plant Chemistry 1957, Hirokawa, Tokyo.

Kjaer, A. (1963). *In* "Chemical Plant Taxonomy" (T. Swain, ed.), pp. 453–473, Academic Press, New York.

Leete, E. (1963). *J. Am. chem. Soc.* **85**, 3523.

Leete, E. (1964). *J. Am. chem. Soc.* **86**, 2509.

Leistner, E. L., and Zenk, M. H. (1968). *Tetrahedron Lett.* 861.

Lewis, I. R., and Warrington, B. H. (1964). *J. chem. Soc.* 5074.

Liebman, A. A., Mundy, B. P., and Rapoport, H. (1967). *J. Am. chem. Soc.* **89**, 664.

Locksley, J., Moore, J., and Scheinmann, F. (1967). *Tetrahedron*, **27**, 2229.

Mabry, T. J. (1966). *In* "Comparative Phytochemistry" (T. Swain, ed.), pp. 231–244, Academic Press, New York.

Mentzer, C., and Fatianoff, O. (1964). "Actualités de Phytochimie Fondamentale", Masson et Cie, Paris.

Mentzer, C. (ed.) (1966). "Actualités de Phytochimie Fondamentale", Series 2, Masson et Cie, Paris.

Mentzer, C. (ed.) (1968). "Actualités de Phytochimie Fondamentale", Series 3, Masson et Cie, Paris.

Moorhead, P. S., and Kaplan, M. M. (eds.) (1967). "Mathematical Challenges to the Neo-Darwinian Interpretation of Evolution", The Wistar Institute Press, Philadelphia, U.S.A.

Mors, W. B., Magalhães, M. T., and Goldschmidt, O. (1962). *Fortschr. Chem. org. Naturstoffe*, **20**, 159.

Nishimura, K. (1966). 4th International Symposium on the Chemistry of Natural Products, Abstract Book, p. 49.

Odham, G. (1966). *Ark. Kemi*, **25**, 543.

Sandermann, W., and Simatupang, M. H. (1966). *Holz Roh- u. Werkstoff*, **24**, 190.

Sörensen, N. A. (1963). *In* "Chemical Plant Taxonomy" T. Swain, ed.), p. 219, Academic Press, New York.

Swain, T. (ed.) (1963). "Chemical Plant Taxonomy", Academic Press, New York.

Swain, T. (ed.) (1966). "Comparative Phytochemistry", Academic Press, New York.

Tse, A., and Towers, G. H. N. (1967). *Phytochemistry*, **6**, 149.

Weatherston, J. (1967). *Quart. Rev.* **21**, 287.

Wilcox, M. E., Wyler, H., Mabry, T. J., and Dreiding, A. S. (1965). *Helv. chim. Acta*, **48**, 252.

Wolfrom, M. L., and Bhat, H. B. (1965). *Phytochemistry*, **4**, 766.

Woods, M. C., Miura, I., Nakadaira, Y., Terahara, A., Maruyama, M., and Nakanishi, K. (1967). *Tetrahedron Lett.* 321.

CHAPTER 6

Chemical Evidence for the Classification of some Plant Taxa

R. HEGNAUER

Department of Experimental Plant Systematics, University of Leiden, Netherlands

I.	Introduction	121
II.	Phytochemistry and Plant Classification	123
	A. The Centrospermae	124
	B. The Pittosporaceae	124
	C. The Cornaceae and Related Families	126
	D. The Plantaginaceae	131
	E. The Hippuridaceae	133
	F. The Papaveraceae	134
III.	Comparative Studies and Classification of Taxa	134
IV.	Conclusion	135
	Acknowledgements	136
	References	136

I. INTRODUCTION

Plants take up various inorganic nutrients from both the soil and the atmosphere and from these elaborate a large number of organic compounds. Both the organic and inorganic constituents of plants are partly continuously metabolized and partly stored in different ways. When we examine the chemical constituents of a given plant, however, we generally can only identify those compounds which are being accumulated to some extent. Our conclusions as to the presence or absence of a given constituent will always depend on its concentration in the plant's tissues, on our method of analysis and on the amount of plant material used. It is probable that many compounds which occur widely, or even ubiquitously, in plants have been identified in only a limited number of the species investigated because their concentration as a rule is too low.

In certain taxa, however, such constituents are readily identified because they are being stored in large amounts and this property represents a highly characteristic feature which may be a valuable taxonomic character. I have discussed the taxonomic impact of such quantitative aspects of chemical characters on several previous occasions (Hegnauer, 1962, 1963b, 1965, 1966).

I put forward the hypothesis that the accumulation of a given compound is far more important in plant taxonomy than its mere presence in trace amounts. Some botanists (e.g. Cranmer and Turner, 1967) nevertheless believe that a quantitative approach to chemical characters is taxonomically unsound and it seems necessary therefore to summarize some of the facts which I believe compel us to make such a distinction.

In comparative phytochemical work it is not always possible to demonstrate the presence of constituents which occur only in trace amounts. If we analyse small samples, e.g. fragments from herbarium sheets, only the major constituents will be identified. Nevertheless, this approach has proved to be fruitful and is frequently used.

Present-day knowledge has shown that many plant constituents are ubiquitous. However, the majority of these products accumulate in large amounts only in rare cases. Hence, accumulation becomes a character worthy of consideration for its own sake. Examples can be found in the case of silicic acid (e.g. Molisch, 1920), aluminium (e.g. Chenery, 1948, 1949; Webb, 1954), shikimic and quinic acids (e.g. Kinzel and Walland, 1966), isocitric acid (e.g. Crassulaceae, *Aloe*), several sugars (e.g. stachyose and verbascose) and most pentitols and hexitols (e.g. adonitol, mannitol sorbitol; Kandler, 1964; Sakai, 1961).

Progress in the speed and sensitivity of analytical methods has rapidly increased our knowledge of the distribution of the so-called secondary metabolites of plants. It has become increasingly clear that many of them occur with high frequency. In most taxa, however, they are present only as minor constituents. Examples may be found among constituents of essential oils (e.g. pinene, cineol, thymol, eugenol; compare also Scora, 1967), coumarins (e.g. esculetin and scopoletin; cf. Goodwin and Taves, 1950; Andreae, 1952), hydroxycinnamic acids (e.g. caffeic acid and its esters), triterpenes (e.g. cycloartanol derivatives) cyanogenic compounds (cf. Rosenthaler, 1923) and even alkaloids (e.g. nicotine; to the genera having species containing trace amounts of this alkaloid [summary, Hegnauer, 1963] must now be added *Aesculus*, *Bacopa* [= *Herpestis*], *Juglans*, *Prunus*, *Solanum* and *Urtica*) and polyacetylenes (Schulte, 1962; Schulte *et al.* 1964, 1965; Fawcett *et al.* 1965).

Regardless of the apparent wide distribution of such metabolites, taxa which show a pronounced tendency to accumulate them are exceptional. Accumulation of any type of compound may represent an outstanding taxonomic character (e.g. polyacetylenes in Umbelliferae and Araliaceae), whereas their mere presence in minute amounts does not. It has to be realized, moreover, that the presence of trace amounts of constituents in a given taxon may come about in several different ways. The compound may be an intermediate or a minor side-product of a general metabolic pathway. On the other hand, the presence of trace amounts of any given compound may also be the result of a late step in an evolutionary line ending with the complete loss of a given character. The latter type of situation probably occurs in some Polygonaceae (loss of anthraquinones) and Ranunculaceae (loss of isoquinoline alkaloids). It is evident that the minor constituents are taxonomically much more meaningful in this instance than the trace substances which are the result of general plant metabolism. The criticism of Cranmer and Turner (1967) appears to be based on the assumption that alteration of metabolism is the most frequent cause of the presence of trace amounts of secondary metabolites in plant tissues. However, this does not very often appear to be the case.

There are fundamental differences between the synthesis or resorption of trace amounts of any one compound and its storage in large amounts, because accumulation is only possible if the whole plant organization and metabolism of the plant have been adapted to a storage condition; for example, many compounds are accumulated in certain taxa in concentrations which would be toxic to ordinary plants. The fact that we have to draw an arbitrary limit between accumulating and non-accumulating plants is not so serious a drawback for taxonomic purposes. For example, in the case of aluminium, a 0·1 % level is the generally accepted limit for accumulation (Hutchinson, 1943; Chenery, 1948; Webb, 1954). Only a limited number of species can surpass this level and these accumulators are confined to a limited number of taxa. Further insight into the phenomenon of aluminium accumulation might help to solve some taxonomic problems.

At this point it is also perhaps worthwhile recalling that many of the characters used in conventional plant taxonomy are quantitative rather than qualitative in character. All tracheophytes bear leaves; the latter show an immense range of variation in size, type and specific shape. This variation is the expression of different growth patterns. Leaf types (e.g. sessile, short petiolate, petiolate, long petiolate; entire, pinnately lobed, pinnatifid, pinnatisect, pinnate) and leaf shapes (compare Exell, 1960, 1962) have to be defined arbitrarily. It is taxonomically sound to consider pinnate leaves as an important character (e.g. Hallier's and Takhtajan's Pinnatae) and at the same time to disregard the sporadic lobation in a taxon with predominantly entire leaves.

Before passing to my theme proper, I should like to emphasize that statements concerning the presence or absence of individual compounds are to be understood as follows: presence means that the constituents mentioned are accumulated to some extent by the respective taxa and absence means that their presence could not be demonstrated by the analytical methods applied (i.e. paper chromatography of leaf and bark extract of herbarium specimens as far as my own results are concerned).

Names and delimitations of orders and families are those of Engler's Syllabus (1964 edition) if not otherwise stated.

II. PHYTOCHEMISTRY AND PLANT CLASSIFICATION

Sixty years ago, Rosenthaler (1907) pointed out that if phytochemistry is to be considered as a serious auxiliary science by plant taxonomists it will have to demonstrate convincingly that criteria derived from comparative chemical studies can be used successfully in plant taxonomy. This postulate is still valid and it is the purpose of the following discussion to demonstrate that chemical characters indeed can be valuable guides for the classification of plant taxa.

A. THE CENTROSPERMAE

The best known recent contribution of phytochemistry to classification concerns betacyanins and betaxanthins (plant pigments known collectively as betalains or chromoalkaloids). Their uniform presence in most families of the Centrospermae and in Cactaceae and Didiereaceae is a strong argument in favour of the hypothesis that these taxa form a natural assembly. Mabry and Turner (1964) showed that *Batis*, a genus highly *incertae sedis*, does not contain these pigments. This is an argument against the association of *Batis* with Chenopodiaceae. However, this example demonstrates clearly that it is very difficult to establish a natural classification of plants and this can only be achieved by comparing all available characters. So far, betalains have not been found in members of either the Molluginaceae (= Aizoaceae–Molluginoideae) or Caryophyllaceae, two families included by most authors on morphological grounds in the Centrospermae. Hence the lack of betalains in *Batis* is not a very strong argument for excluding this genus from the Centrospermae. On the other hand, the presence of betalains in the Didiereaceae undoubtedly supports the inclusion of this family in the order. The classification of those taxa mentioned which lack betalains must be settled by continuing to study the whole gamut of their characters in a comparative manner. As far as other chemical constituents are concerned the Molluginaceae and Caryophyllaceae fit rather well into the pattern of the order Centrospermae as understood by most modern taxonomists, i.e. the totality of the chemical characters known at present points to a close centrospermous affinity, despite the lack of betalains.

B. THE PITTOSPORACEAE

The Pittosporaceae are trees, shrubs or woody climbers with exstipulate, generally entire and evergreen leaves and with hypogynous, tetracyclic flowers with 5 sepals, petals, and stamens and usually 3–5 carpels united in a unicellular, single-styled ovary with 2–5 parietal placentae. The fruits are many-seeded capsules or berries, and the seeds contain a small embryo embedded in copious endosperm. The family is restricted to the warmer regions of the Old World. Of its nine genera, six are endemic in Australia; *Hymenophyllum* and *Citrio-batus* occur in North Australia and Malaysia and *Pittosporum* is widely distributed in Africa, Australia, Asia, Malaysia, New Zealand and other Pacific Islands. In most modern systems (Pritzel, 1930; Wettstein, 1935; Gundersen, 1950; Pulle, 1952; Cronquist, 1955; Takhtajan, 1959; Emberger, 1960; Engler, 1964; Airy Shaw, 1966) Pittosporaceae are classified in the vicinity of woody

Saxifragaceae (especially Escallonioideae). Hutchinson (1959) classifies his Pittosporales next to Bixales but remarks that they "might be equally well placed near the Cunoniales".

When examining the chemical characters of Pittosporaceae it became clear to me that the family is chemically unlike members of the saxifragaceous stock. Observations reported in the literature make it possible to summarize phytochemical features of Pittosporaceae as follows:

(a) Saponins, with α- and β-amyrin type triterpenes, occur frequently.

(b) All members produce essential oils; some of their outstanding components are anthranilic acid, indole, decanal and the furanocoumarin, bergaptene.

(c) Caffeic acid, ferulic acid and the flavonols kaempferol and quercetin, are almost ubiquitous.

(d) Leucoanthocyanins are very rare and, if present, represented by leucocyanidin only.

(e) This picture may be completed by negative, but taxonomically by no means less important, evidence: ellagic acid, flavonoid compounds with a trihydroxylated B-ring, true tannins and cyanogenic glycosides have never been found in the species investigated.

A search of the literature revealed that as early as 1884 van Tieghem suggested that Pittosporaceae are most probably rather closely related to Araliaceae and Umbelliferae; they possess schizogenous oil ducts and some specific features of root anatomy (schizogenous ducts located in such a manner in the pericycle that a characteristic arrangement of adventitious roots results) in common with the two families mentioned. Later Schürhoff (1929) investigated the embryology of *Pittosporum*; he observed trinucleate pollen grains, tenuinucellate ovules with one integument and nuclear endosperm, and concluded that Pittosporaceae should be placed as proposed by van Tieghem. According to Metcalfe and Chalk (1950) the wood anatomy (e.g., septate fibres; paratracheal parenchyma) of the family agrees much better with van Tieghem's proposal than with the generally accepted saxifragaceous affinity. Huber (1963) arrives at similar conclusions. These arguments notwithstanding, taxonomists unanimously have continued to ally Pittosporaceae with the saxifragaceous stock mostly on account of flower morphology. According to Eames and MacDaniels (1947) the inferior ovary of Araliaceae originated by adnation of sepals, petals and filaments. If, in fact, this is the way Araliaceae acquired epigyny, the hypogynous flower of Pittosporaceae with its sometimes connate calyx and its petals often coherent in the basal part, can be interpreted as an initial stage preceding adnation. Araliaceae and Umbelliferae have mostly dissected leaves, whereas Pittosporaceae usually bear simple leaves. There are, however, species of *Pittosporum* with highly polymorphic leaves. Seedling leaves and juvenile leaves may be pinnatifid whilst adult leaves are entire (Laing and Gourlay, 1935).

5*

It therefore seemed worthwhile to examine the chemical characters further. We looked at leaves of four species of *Pittosporum* for iridoid glucosides (present in *Escallonia* and *Hydrangea*), and gallic and ellagic acids (present in many of the Rosales). The results obtained, compared with those for *Hedera helix* (Araliaceae) are shown in Table I.

The results agree with the hypothesis that the Pittosporaceae are related to the Araliaceae. This hypothesis will be strengthened by further investigations of Pittosporaceae for other chemical markers typical of Araliaceae and Umbelliferae. For example, Professor Bohlmann (Techn. Univ. Berlin) has isolated two polyacetylenes from *Pittosporum buchananii* very similar to those occurring in Araliaceae and Umbelliferae (Bohlmann and Rode, 1968).

TABLE I

Leaf constituents of four species of *Pittosporum* compared with *Hedera Helix*

Species	Iridoid glucosides	Gallic acid	Ellagic acid	Chlorogenic acid	Quinic acid
Pittosporum tenuifolium Soland. ex Gaertn. (= *P. mayi* Hort.): EPL 9831[a]	0	0	0	+	+
P. tobira Ait.: EPL 9830[a]	0	0	0	+	+
P. undulatum Vent.: EPL 9832[a]	0	0	0	+	+
P. viridiflorum Sims: EPL 9837[a]	0	0	0	+	+
Hedera helix L.: EPL 11699	0	0	0	+	+

[a] Herbarium numbers (Lab. Experimentele Plantensystematiek, Leiden). Herbarium material used, except in the case of *Hedera*.

Taxonomists should take note of the anatomical, embryological and chemical evidence summarized here and should seriously reconsider the classification of Pittosporaceae. Inclusion of the family in Araliales (sensu Takhtajan; i.e. comprising Araliaceae and Umbelliferae only) appears much more likely to reflect its natural affinities than does its still generally accepted incorporation in the Rosales.

C. THE CORNACEAE AND RELATED FAMILIES

Takhtajan united in the order Cornales, the families Cornaceae, Garryaceae, Davidiaceae, Nyssaceae, Alangiaceae, Mastixiaceae and Torricelliaceae. In his system Cornales and Araliales together form the "Überordnung" Umbelliflorae (identical with the order Umbelliflorae of Engler's Syllabus), which is supposed to have evolved from Rosiflorae (Cunoniaceae in particular). A similar hypothesis with regard to the origin of Cornales was put forward by

Cronquist (1955), although he does not, however, accept a close relationship between Cornales and Araliales (called Umbellales by Cronquist) but derives the latter from the rutaceous alliance (i.e. Sapindales *sensu lato*). Except for Cronquist, most modern authors (e.g., Faure [1924; exclusive of Garryaceae], Wettstein, Gundersen, Pulle, Rodriguez [1957], Takhtajan, Emberger, Engler, Airy Shaw: sub Helwingiaceae; Philipson [1967]) assume that the Cornaceae and Araliaceae are closely related and that this is a more satisfying classification than others which have been put forward (Hoare, 1915; Metcalfe and Chalk, 1950; Huber, 1963).

We became interested in this problem of plant taxonomy because the chemistry of all species of *Cornus* investigated (presence of cornin [= verbenalin], ellagi- and gallitannins, leucoanthocyanins; lack of petroselinic acid in seed oils and apparent absence of polyacetylenic compounds [Yosioka *et al.* 1966]) appears to be very different from the chemistry of Araliaceae (Hegnauer, 1965). Cornales, however, represent a rather heterogenous assembly of genera. Therefore we have started investigations which aim at obtaining more information about the chemistry of as many genera of the Cornales as possible. Hitherto, we looked only for iridoid glycosides and phenolic acids. The information acquired so far is summarized in Table II (Wieffering, 1966; unpublished results of Miss L. H. Fikenscher).

The presence of iridoid compounds (Table II; Fig. 1) forms a chemical link between most of the genera investigated.

In the case of the phenolic acids, however, two main groups appear to exist. The first group, characterized by the presence of gallic and ellagic acids (and corresponding tannins), comprises *Camptotheca*, *Davidia*, *Nyssa*, *Cornus*, *Corokia* and *Mastixia*. The second lacks both gallic and ellagic acids, but contains caffeic acid (chlorogenic acid) as the chief phenolic acid; *Aucuba*, *Garrya*, *Griselinia* and *Helwingia* are members of this group.

The results with regard to *Alangium* and *Helwingia* are inconclusive because no phenolic acids could be identified in the former taxon and none of the iridoid glucosides, aucubin, cornin, loganin, could be detected in *Helwingia*. The spots giving positive reactions to the Kedde and Legal reagents (unknowns in Table II) of course, may indicate cornin-like substances. This question, however, can only be decided by the isolation and identification of the constituents responsible.

The most thorough recent discussions of systematics of Cornaceae *sensu lato* are those of Eyde (1963, 1964, 1966, 1968). Basing his arguments principally on features of flower anatomy, this author suggests several natural groups within the alliance:

(a) Cornaceae proper, characterized by an anomalous vascular pattern of the ovaries, comprise *Cornus*, *Mastixia*, *Torricellia*, Alganiaceae (*Alangium*) and Nyssaceae (*Camptotheca*, *Davidia*, *Nyssa*).

(b) The *Aucuba*, *Garrya*, *Griselinia* group, which have unilocular ovaries,

TABLE II

Distribution of iridoid glucosides and some phenolic acids in members of Cornales sensu Takhtajan

Taxon	Plant part[a]	Iridoid glucosides[b]	Gallic acid[c]	Ellagic acid[c]	Caffeic acid[c]	Quinic acid[d]
Alangiaceae						
Alangium rotundifolium (Hassk.) Bloemb.	L	} Cornin; unknown 0·25 and 0·76	0	0	0	0
EPL 8212	T		0	0	0	0
Alangium spec. EPL 8205, 8206, 8207, 8208	L		0	0	0	−
Garryaceae						
Garrya elliptica Dougl.	L	Aucubin, unknown 0·65, 0·53 and 0·33	0	0	Ch	+
EPL 11619						
Nyssaceae						
Camptotheca acuminata Decne.						
EPL 8338	T	0	+	+	+	−
EPL 8339	L	0	(+)	+	+	−
Davidia involucrata Baill.						
EPL 8752	L, T	0	+	+	0	+
Nyssa javanica (Bl.) Wang.	L	Cornin	+	+	+	−
EPL 8197	T	Cornin	0	0	+	−
Nyssa silvatica Marsh., EPL 8751	L	Cornin	−	−	−	−
(living plant No. 66690)	T	Cornin	−	−	−	−

Cornaceae

Taxon	Part[a]	Compounds[b]				
Aucuba japonica Thunb., EPL 8753	L, T	Aucubin	0	0	Ch	0
Cornus mas L., EPL 4827	L	—	+	+	+	−
Corokia cotoneaster Raoul EPL 5821; 11620	L, T	Cornin	GT	−	0	+
Corokia virgata Turrill, EPL 9519	L, T	Cornin	+	0	0	+
Griselinia littoralis Raoul EPL 4085; 8754	L, T	Cornin	0	0	Ch; IsoCh	+
Helwingia japonica Dietr. EPL 9518	L	Unknown 0·76 verbenalic acid?	0	0	+; Ch	0
Mastixia arborea (Wight) C.B. Clarke, EPL 5084	L	Loganin; loganic acid	GT	−	Ch	+
	B	Loganin; loganic acid; cornin?	−	−	−	−
Mastixia pentandra Blume EPL 7660	L	Verbenalic acid; verbenalol?	+	+	Ch	+
Mastixia rostrata Blume EPL 8204	L	Cornin; unknown 0·76?	+; GT	+	(+)	+
EPL 8202	T	Cornin; loganin; loganic acid	−	−	−	−
EPL 8203	L	—	+	+	+	−
	L	—	+	+	+	−
Mastixia trichotoma Blume EPL 8200	B	Cornin	+; GT	+	+	−
	L	—	0	+	+	+
Mastixia spec. EPL 8198	L	—	+	+	(+)	−
Mastixia spec. EPL 8199	L	—	+	(+)	+	−

+, present; 0, not present; −, not yet tested.

[a] L = leaf; T = twig; B = bark. Paper chromatographic analysis of extracts from herbarium specimens (from the Laboratory of Experimental Plant Systematics, Leiden).

[b] Experimental details (see Wieffering, 1966); unknown spots symbolized by R_f values in BAW, Cornin is mostly accompanied in all species examined by verbenalic acid, and still other compounds (e.g., verbenalol) probably derived from, or related to, Cornin (= verbenalin).

[c] In hydrolysed extracts; gallotannins (GT), Chlorogenic (Ch) and Isochlorogenic (IsoCh) acid in unhydrolysed extracts.

[d] In unhydrolysed extracts.

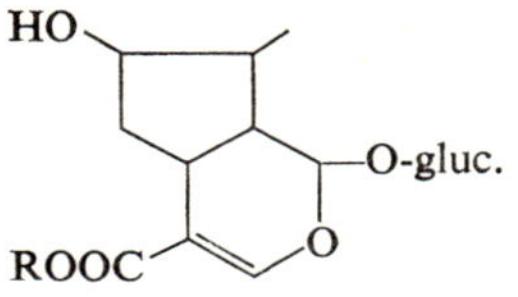

(I) R = CH₃: Loganin
    R = H: Loganic acid } Mastixia

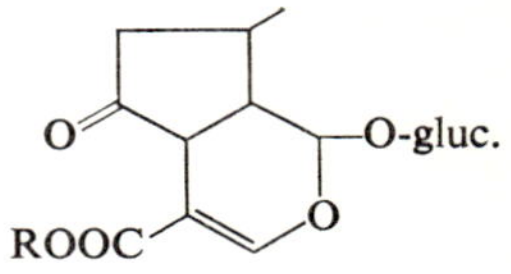

(II) R = CH₃: Verbenalin (= Cornin) } Mastixia
     R = H: Verbenalic acid   Nyssa
                                       Griselinia
                                       Corokia
                                       Alangium

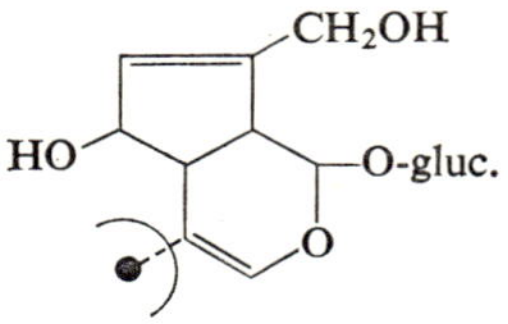

(III) Aucubin } Aucuba
              Garrya

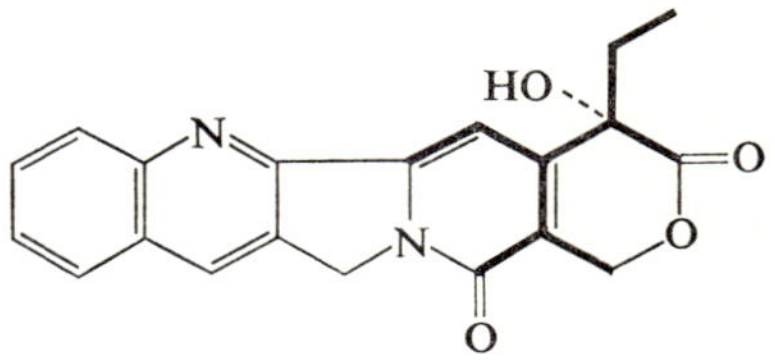

(IV) Camptothecin: Camptotheca

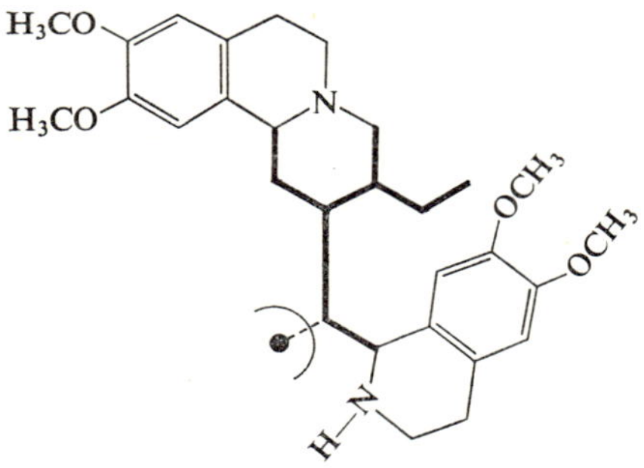

(V) Emetine (and related alkaloids):-Alangium

FIG. 1. Constituents of presumed iridoid origin present in Cornales.

is assumed to have had a common origin with Cornaceae proper. A relationship between *Aucuba* and *Griselinia* is also accepted by Kubitzki (1963), who has worked along the same lines as Eyde. Philipson (1967) pointed out that *Griselinia* shows very complex interrelationships; it resembles Cornaceae, Escalloniaceae and Araliaceae, each in some respects.

(c) *Corokia* is presumed to have *Argophyllum*, a genus commonly placed in Saxifragaceae–Escallonioideae, as its nearest relative. Huber (1963, p. 21), on the other hand, believes that *Corokia* is a true member of Cornaceae.

(d) *Helwingia* is suggested to be a member of Araliaceae.

(e) *Curtisia, Kaliphora* and *Melanophylla* are supposed to have no affinities with Cornaceae proper.

At the same time Eyde (1968) points out that floral anatomy does not favour incorporation of Cornaceae and Nyssaceae in Umbelliflorae.

Chemical data (as far as current information is available) appears to coincide rather well with these conclusions. The Cornaceae proper contain gallo- and ellagitannins and iridoid compounds. Their chemistry suggests that they are closer to some members of Rosales (gallic and ellagic acid are rather common in this order; loganin occurs in *Hydrangea*, asperuloside in *Escallonia* and monotropeoside in *Liquidambar*) than they are to Umbelliflorae. The *Aucuba*, *Garrya*, *Griselinia* group (no accumulation of gallic and ellagic acids; production of iridoid glucosides retained) offers no difficulties from the phytochemical point of view. These genera might indeed represent derivatives from the cornaceous stock. *Griselinia*, however, appears to stand a little apart by containing large amounts of saponins with pentacyclic triterpenes (A_1-barrigenol, oleanolic acid: Silva-Balocchi, 1968) as their aglycones. *Griselinia* might thus equally well stand nearer to some representatives of the rosaceous stock than to members of the cornaceous stock.

Corokia, like Cornaceae proper, shows a combination of chemical characters not uncommon in the rosaceous stock. Chemically, it might be classified with Cornaceae proper or with some members of Rosales.

In conclusion, we may state that the chemistry of most members of Cornales definitely points to their exclusion from Umbelliflorae and suggests an affiliation with some members of Rosales. Removal of Cornales from Umbelliflorae and inclusion of Pittosporaceae in the latter order requires, and makes possible, reconsideration of the relationships of the Umbelliflorae defined in the above manner (i.e. comprising Pittosporaceae, Araliaceae and Umbelliflorae). There are striking chemical resemblances between members of Rutales and Umbelliflorae. The chemistry of *Pittosporum* is reminiscent to some degree, of Rutaceae (essential oils, with anthranilic acid and bergaptene) and of Araliaceae (wealth of triterpenic saponins; acetylenic compounds). I suggest that Pittosporaceae represent in some way an intermediate stage in the evolution of Araliaceae and Umbelliferae from the rutaceous stock. Cornaceae, on the other hand, are closer in their chemistry to some members of Rosales or to the Gentianales (Rubiaceae, e.g.) and Dipsacales (Caprifoliaceae, e.g.). The Cornaceae might thus represent in several respects intermediary stages between polypetalous perigynous Rosales and epigynous members of sympetalous dicotyledons (Fig. 2).

This hypothesis has evidently to be tested thoroughly by comprehensive comparative studies of many other types of characters of the taxa concerned.

D. THE PLANTAGINACEAE

In modern systems the Plantaginaceae are placed in the monotypic order Plantaginales which is derived from Tubiflorae (Cronquist, Engler's Syllabus) or from Primulales (Gundersen, Hutchinson). Wettstein, Takhtajan and

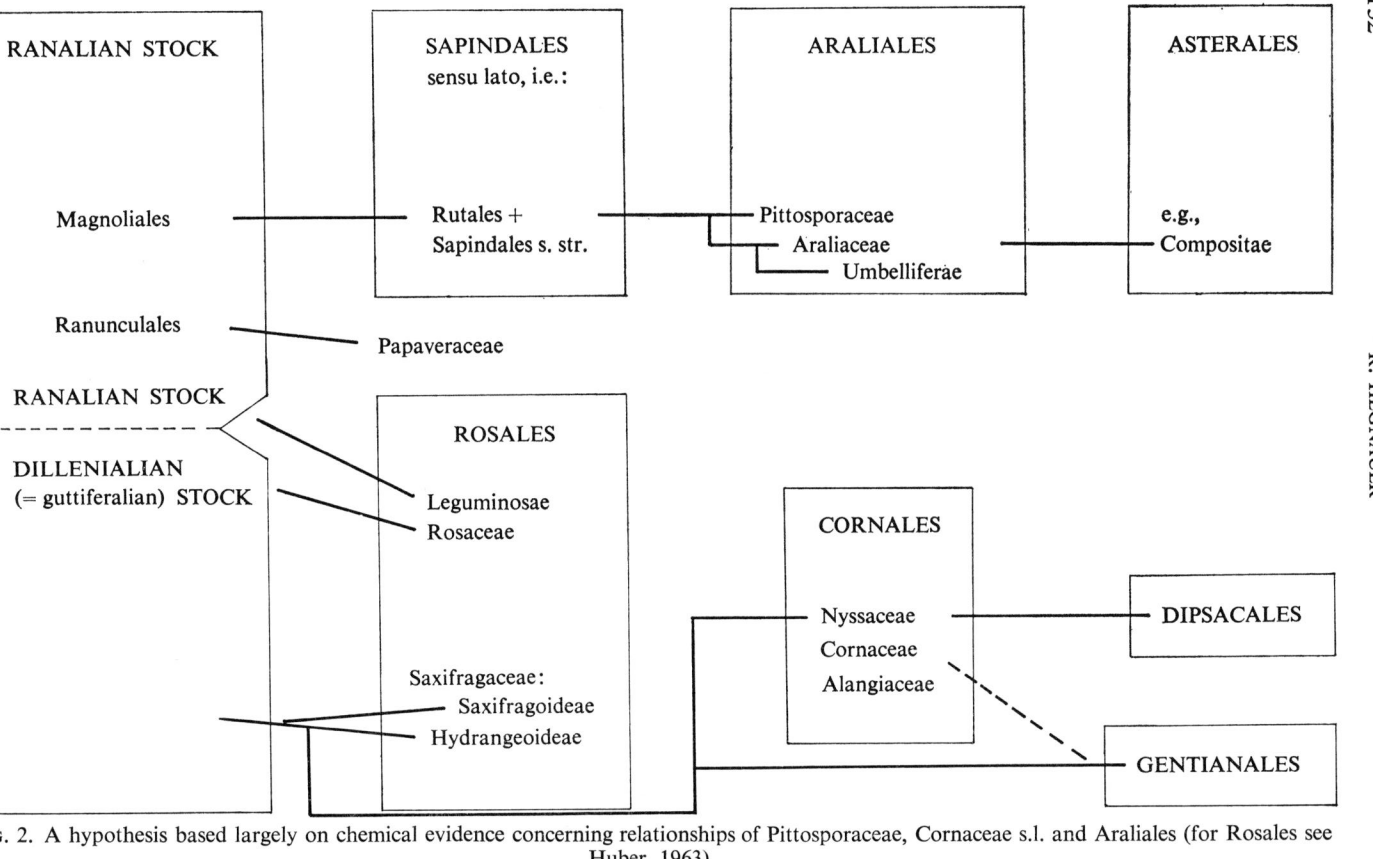

FIG. 2. A hypothesis based largely on chemical evidence concerning relationships of Pittosporaceae, Cornaceae s.l. and Araliales (for Rosales see Huber, 1963).

Emberger include Plantaginaceae in Tubiflorae. *Plantago,* the chief genus of the family, has been chemically well studied; iridoid glucosides (aucubin, catalpol, methylcatalpol), supposedly iridoid alkaloids (e.g. plantagonin, indicamin), glycosides (plantaginin and homoplantaginin) of the flavones scutellarein and dinatin, free pentacyclic triterpene carboxylic acids (oleanolic and ursolic acid) have been detected in one or more species. Stachyose-type oligosaccharides are stored in perennial roots and rhizomes, hexitols (e.g. sorbitol) in leaves and planteose, fatty oil and reserve celluloses in seeds. We have examined *Litorella uniflora* (L.) Aschers. (EPL number 11358) , a wholly aquatic member of the family, for iridoid compounds and identified aucubin. As Bourdu *et al.* (1963) have already shown that the vegetative parts of *Litorella* store stachyose and that its seeds contain planteose, it is evident that adaptation to aquatic life has not much changed metabolism.

The Plantaginaceae are chemically rather homogeneous and at the same time very similar to several members of Tubiflorae. On the other hand fundamental chemical differences exist between Plantaginaceae and Primulales. Plantaginaceae can be interpreted as an anemogamous member of Tubiflorae with rather clear affinities to Scrophulariaceae. Evidently the family deviates from the majority of Tubiflorae by the tetramerous, anemogamous flowers. This, however, is by no means a character which necessitates the introduction of a separate order for Plantaginaceae. If, for instance, Callitrichaceae are included in Tubiflorae as in the most recent edition of Engler's Syllabus, a classification which is in excellent agreement with chemical evidence (Wieffering, 1966; Hegnauer, 1967), it would only be logical not to place Plataginaceae in an order of their own because they deviate less from the general plan of Tubiflorae than do Callitrichaceae.

E. THE HIPPURIDACEAE

The Hippuridaceae comprise only the monotypic genus *Hippuris* with the polytypic, aquatic species *Hippuris vulgaris* L. In most modern systems (except Pulle's: Hippuridales derived from Solanales [i.e. Tubiflorae]) this taxon is included in Haloragaceae (Gundersen, Hutchinson) or placed in a monotypic family classified in the neighbourhood of Haloragaceae (Wettstein, Cronquist, Takhtajan, Emberger, Engler's Syllabus, Airy Shaw).

The chemical characters of *Hippuris* are, however, very different from those of Haloragaceae. Therefore, reconsideration of this taxon with extremely reduced flowers is desirable. *Hippuris* stores stachyose and produces both aucubin and catalpol. Unlike Haloragaceae and related families, it does not contain leucoanthocyanins, flavonols and hydrolysable tannins. In the light of chemical evidence, Pulle's suggestion is the only tenable hypothesis for the classification of Hippuridaceae. A sympetalous affinity had already been considered possible by Juel (1911) on account of embryological features. Like *Callitriche, Hippuris* is most probably a wholly aquatic offshoot of Tubiflorae which might have its nearest relatives in Plantaginaceae (compare *Litorella*).

The examples of *Litorella* and *Callitriche* demonstrate that evolution to aquatic life which involves far reaching alterations in flower morphology (*Callitriche*) and general anatomy may leave metabolic patterns relatively unaltered. Hence the chemical characters of *Hippuris*, a taxon with extremely reduced flowers, become important guides for classification.

F. THE PAPAVERACEAE

The classification of Papaveraceae is still controversial. Seven years ago I stressed that chemistry of the constituents suggest that the family is out of place in the order Rhoeadales as usually accepted (Hegnauer, 1961). Long ago Hallier (1912) proposed that a close relationship existed between Papaveraceae and the ranalian stock. At the same time he pointed to fundamental differences between Papaveraceae and other families (e.g., Capparidaceae, Cruciferae) usually included in the same order. His ideas were further developed by Takhtajan, who restricted his order Papaverales (derived from Ranales) to Papaveraceae and Fumariaceae. For his Capparidales (= Rhoeadales less Papaveraceae), Takhtajan assumed relationships with Cistales (e.g., Flacourtiaceae). As far as data are available, phytochemistry suggests that this is a more natural classification for the taxa concerned. The strong chemical evidence in favour of Hallier's and Takhtajan's view summarized in 1961 appears to have stimulated a reinvestigation of serological (Frohne, 1962) and morphological (Merxmüller and Leins, 1967) properties of the families. The results agreed wholly with chemical evidence. It is to be hoped that general agreement between taxonomists will be reached soon with regard to the classification of the families mentioned.

III. COMPARATIVE STUDIES AND CLASSIFICATION OF TAXA

From the preceeding discussion of relationships between and within the Cornaceae, Umbelliflorae and Rhoeadales, one cardinal point of classification of larger taxa emerges. A natural classification of families and orders only becomes possible after careful consideration of the naturalness of the taxa to be classified. If, for instance, Rhoeadales are derived from Ranunculales as in Engler's Syllabus or from Guttiferales (parallel to Bixales) by Cronquist, both views have their merits. In the Syllabus, the presumed affinities are based largely on Papaveraceae and in Cronquist's system on Capparidaceae and Cruciferae. Both authors, however, do not seem to have paid attention to the following essential question: Do the Rhoeadales represent a natural taxon? After an embryological study of *Garrya*, Kapil and Mohana Rao (1966) concluded "therefore the assignment of the Garryaceae to the order Umbelliflorae appears quite satisfactory". They compared *Garrya* with Alangiaceae and Cornaceae accepting both as true members of Umbelliflorae. Hence their statement is justified only as far as the three taxa mentioned are concerned.

At the same time their conclusion is inaccurate and misleading because it suggests a close relationship between *Garrya*, Araliaceae and Umbelliferae.

My own hypothesis concerning relationships of Cornales (p. 131) can also be used to illustrate this most important point. After careful comparisons of many characters, Huber (1963) arrived at the conclusion that the order Rosales, as it is commonly circumscribed, represents a very unnatural taxon. He suggested the following rearrangements and relationships.

Cornales: Comprising Philadelphaceae, Styracaceae, Symplocaceae, Escalloniaceae, Diapensiaceae, Aquifoliaceae and Cornaceae. Intimately related with Ericales; also related with Saxifragales (sensu Huber).

Saxifragales: Comprising Grossulariaceae, Crassulaceae, Penthoraceae, Saxifragaceae *sensu stricto* and most probably *Gunnera* and *Cephalotus*. Related with Haloragidales and Araliales and less intimately with Cornales.

Pittosporales: Comprising Pittosporaceae and probably Byblidaceae. Related with Araliales.

Hamamelidales: Comprising Altingiaceae, Hamamelidaceae (excluding *Rhodoleia*) and Platanaceae.

Rosales: Comprising Rosaceae *sensu stricto*, Mespilaceae and Amygdalaceae. Relationships obscure.

Leguminosales: Comprising Mimosaceae, Caesalpiniaceae and Papilionaceae. Affinities with Sapindales and less so with Myrtales.

In the light of the suggestions of Huber it is evident that a hypothesis deriving Cornales from Rosales is meaningless if the members of Rosales to which affinities are supposed to exist, are not clearly specified. This is the more important because according to Huber affinities of Cunoniaceae are still completely obscure (compare Cunoniales of Takhtajan!). Figure 2 is an attempt to illustrate my own view concerning relationships of Cornales. Some of the ideas put forward by Huber are very attractive in the light of present-day chemical evidence. Many more comparative studies are needed, however, before well founded conclusions can be reached.

In order to ultimately put forward as natural as possible a system of plants we still urgently need intensive comparative studies within the commonly accepted taxa. This is especially true at family and order level. Many of our families and orders represent rather unnatural entities. Consequently, in many instances, the classification at higher levels (families within orders; interrelationships of orders), becomes highly uncertain and mainly speculative. Comparative phytochemistry is one of the tools at our disposal to verify naturalness of morphologically defined and commonly accepted taxa. In my opinion this represents one of the major tasks of chemotaxonomy today.

IV. CONCLUSION

I hope to have been able to demonstrate that comparative phytochemistry can be a very valuable tool in the hands of plant taxonomists. This was anticipated long ago by botanists like A. P. De Candolle, H. Hoffmann (1846), H. Hallier, H. Molisch and others, and by phytochemists like Rochleder, M. Greshoff, H. Thoms, L. Rosenthaler and R. Jaretzky. It is, however, only in our days that methods of analysis and structural elucidation of compounds have reached a stage which makes large scale surveys of the chemistry of plants possible. This, evidently, is essential for a sound taxonomic interpretation of chemical characters. It is not strange, therefore, that the scepticism shared by most classical taxonomists against phytochemistry as a taxonomic tool has only recently begun to diminish (compare, e.g., Merxmüller, 1963, 1967; Ehrendorfer, 1965, 1967; and section Chemotaxonomia in Excerpta Botanica).

ACKNOWLEDGEMENTS

I am most indebted to Dr W. Meijer (Forest Botanist, Sandakan) for material of *Alangium*, *Mastixia* and *Nyssa javanica*; to Dr R. H. Eyde (Smithsonian Institution, Washington) for material of *Camptotheca*, for a copy of a manuscript and for stimulating discussions; to Professor F. Bohlmann (Techn. Univ. Berlin) for kind co-operation in looking for acetylenic compounds in *Pittosporum* and for the communication of results; to Professor H. Gloor (Genetical Institute, Univ. of Leiden) for corrections of the English text; and to Dr Lucie H. Fikenscher (Lab. Experiment. Plantensystematiek, Leiden) for experimental work.

REFERENCES

Airy Shaw, H. K. (1966). *In* "A dictionary of flowering plants and ferns" (J. C. Willis), 7th ed., Univ. Press, Cambridge.
Andreae, W. A. (1952). *Nature*, **170**, 83.
Bohlmann, F., and Rode, K. M. (1968). *Chem. Ber.* **101**, 1829.
Bourdu, R., Cartier, D., and Gorenflot, R. (1963). *Bull. Soc. bot. Fr.* **110**, 107.
Chenery, E. M. (1948). *Kew Bull.* **1948**, 173
Chenery, E. M. (1949). *Kew Bull.* **1949**, 463.
Cranmer, M. F., and Turner, B. (1967). *Evolution*, **21**, 508.
Cronquist, A. (1955). *Bull. Jard. bot. État, Brux.* **27**, 13.
Eames, A. J., and MacDaniels, L. H. (1947). "An introduction to plant anatomy", McGraw-Hill, New York.
Ehrendorfer, R. (1965). *Fortschr. Bot.* **27**, 373–376.
Emberger, L. (1960). "Traité de botanique systématique. II Les végétaux vasculaires", Fasc. 1, pp. 672–694, Masson, Paris.
Engler, A. (1964). "Syllabus der Pflanzenfamilien", 12. Aufl., Bd. 2, Gebr. Bornträger, Berlin.
Exell, A. W. (1960). *Taxon*, **9**, 245.
Exell, A. W. (1962). *Taxon*, **11**, 145, 245.
Eyde, R. H. (1963). *J. Arnold Arbor.* **44**, 1.
Eyde, R. H. (1964). *Am. J. Bot.* **51**, 1083.
Eyde, R. H. (1966). *Am. J. Bot.* **53**, 833.
Eyde, R. H. (1968). *Phytomorphology*, **17** (Maheshwari Memorial Volume). In press.
Faure, A. (1924). "Etude organographique, anatomique et pharmacologique de la famille des Cornacées (groupe des *Cornales*)", Imprimérie Centrale du Nord, Lille.
Fawcett, C. H., Spencer, D. H., Wain, R. L., Jones, E. R. H., LeQuan, M., Page, C. B., and Thaller, V. (1965). *Chem. Commun.* **1965**, 422.

Frohne, D. (1962). *Planta Medica*, **10**, 283.

Goodwin, H., and Taves, C. (1950). *Am. J. Bot.* **37**, 224.

Gundersen, A. (1950). "Families of Dicotyledons", Chronica Botanica Comp., Waltham, Mass.

Hallier, H. (1912). *Arch. Sci. Neerl. Sci. Exactes Nat.*, Série IIIB, **1**, 146.

Hegnauer, R. (1962). "Chemotaxonomie der Pflanzen", Vol. I, p. 164, Birkhäuser, Basel.

Hegnauer, R. (1963a). "Chemotaxonomie der Pflanzen", Vol. II. Birkhäuser, Basel.

Hegnauer, R. (1964). "Chemotaxonomie der Pflanzen", Vol. III. Birkhäuser, Basel.

Hegnauer, R. (1966). "Chemotaxonomie der Pflanzen", Vol. IV. Birkhäuser, Basel.

Hegnauer, R. (1966). "Chemotaxonomie der Pflanzen", Vol. IV.

Hegnauer, R. (1961). *Planta Medica*, **9**, 37.

Hegnauer, R. (1963a). *In* "Chemical Plant Taxonomy" (T. Swain, ed.), pp. 397–399, Academic Press, London and New York.

Hegnauer, R. (1965a). *Lloydia*, **28**, 267.

Hegnauer, R. (1965b). *In* "Beiträge zur Biochemie und Physiologie von Naturstoffen"; K. Mothes, p. 467, G. Fischer, Jena.

Hegnauer, R. (1967). *Pure appl. Chem.* **14**, 173.

Hoar, C. S. (1915). *Ann. Bot.* **29**, 55.

Hoffmann, H. (1846). "Schilderung der deutschen Pflanzenfamilien vom botanisch-descriptiven und physiologischchemischen Standpunkte", G. F. Heyer's Verlag, Giessen.

Huber, H. (1963). *Mitt. bot. Staatamml. Münch.* **5**, 1.

Hutchinson, G. E. (1943). *Q. Rev. Biol.* **18**, 1, 128, 242, 331.

Hutchinson, J. (1959). "The families of flowering plants", Vol. 1, second ed., Clarendon Press, Oxford.

Juel, H. O. (1911). *Nova Acta Regiae Soc. Sci. Upsaliensis*, Ser. IV, **2**, fasc. 2, 1–26.

Kandler, O. (1964). *Ber. dt. bot. Ges.* **77**, Sondernummer, 1. Generalversammlungsheft 62–73.

Kapil, R. N., and Mohana Rao, P. R. (1966). *Phytomorphology*, **16**, 564.

Kinzel, H., and Walland, A. (1966). *Z. Pflphysiol.* **54**, 371.

Kubitzki, K. (1963). *Ber. dt. bot. Ges.* **76**, 33.

Laing, M., and Gourlay, H. W. (1935). *Trans. R. Soc. New Zeal.* **65**, 44.

Mabry, T. J., and Turner, B. L. (1964). *Taxon*, **13**, 197.

Merxmüller, H. (1963). *Fortschr. Bot.* **25**, 93–97.

Merxmüller, H. (1967). *Ber. dt. bot. Ges.* **80**, 608.

Merxmüller, H., and Leins, P. (1967). *Bot. Jb.* **86**, 113.

Metcalfe, C. R., and Chalk, L. (1950). "Anatomy of the Dicotyledons", Clarendon Press, Oxford.

Molisch, H. (1920). *Sitz. Akad. Wiss. Wien, Math.-naturw. Kl.*, Abt. 1, **129**, 261.

Philipson, W. R. (1967). *N. Z. J. Bot.* **5**, 134.

Pritzel, E. (1930). "Natürl. Pflanzenfamilien", 2. Aufl., Bd. 18a, Leipzig.

Pulle, A. A. (1952). "Compendium van de terminologie, nomenclatuur en systematiek der zaadplanten", third ed., Oosthoek, Utrecht.

Rodriguez, R. L. (1957). *Univ. of California, Publications in Botany*, **29**, 145.

Rosenthaler, L. (1907). *Beih. bot. Zbl.* **21**, Abt. 1, 304.

Rosenthaler, L. (1923). *Biochem. Z.* **134**, 215.

Sakai, A. (1961). *Nature*, **189**, 416.

Scora, R. W. (1967). *Taxon*, **16**, 499.

Schulte, K. E. (1962). "XXI° Congresso di Sci. Farm., Pisa 1961; Conferenze i Commun.", Ed. Fed. Ordini Farmacisti Italiani, Roma, pp. 798–308.

Schulte, K. E., Reisch, J., and Bornfleth, H. (1964). *Arch. Pharm.* **297**, 443; *Arzneimittelforsch.* **14**, 844.

Schulte, K. E., Reisch, J., and Rheinbay, J. (1965). *Phytochemistry*, **4**, 481.

Schürhoff, N. N. (1929). *Beitr. Biol. Pfl.* **17**, 72.
Silva, M., and Balocchi, M. (1968). *Phytochemistry*, **7**, 333.
Takhtajan, A. (1959). "Die Evolution der Angiospermen", G. Fischer, Jena.
Tieghem, van, Ph. (1884). *Bull. Soc. bot. Fr.* **31**, 383.
Webb, L. J. (1954). *Aust. J. Bot.* **2**, 176.
Wettstein, R. (1935). "Handbuch der systematischen Botanik", 4. Aufl., F. Deuticke, Wien.
Wieffering, J. H. (1966). *Phytochemistry*, **5**, 1053.
Yosioka, I., Kimura, T., Imagawa, H., and Takara, K. (1966). *J. Pharm. Soc. Japan*, **86**, 1216.

CHAPTER 7

Chemotaxonomy of the Sesquiterpenoids of the Compositae

V. HEROUT AND F. ŠORM

Czechoslovak Academy of Sciences, Prague, Czechoslovakia

I. Introduction 139
II. The Definition of Sesquiterpene Lactones as Taxonomic Characters . 140
III. The Probable Biosynthetic Pathway to Sesquiterpene Lactones . . 142
IV. Sesquiterpene Lactones and the Classification of the Compositae into
Tribes 144
V. Sesquiterpene Lactones and the Classification within Tribes . . 154
    A. Cynareae 154
    B. Anthemideae 156
    C. Heliantheae and Helenieae 156
    D. Senecioneae 158
VI. Limitations in the Taxonomic Significance of the Phytochemical Data . 161
VII. Conclusion 162
References 163

I. Introduction

The Compositae (Asteraceae) contain a very large number of species and is, perhaps, the largest of all plant families; this makes the systematic classification rather difficult. It is true that the individual tribes, as defined in the system of Cassini or Bentham and later by Hoffmann (1894), represent a very natural classification. However, it is still very difficult to determine which sub-unit of the Compositae represents the basis, and which are derived—i.e. which sub-units are phylogenetically young. The efforts of many specialists to settle these questions have been summarized by Cronquist (1955). However, definite conclusions on the phylogeny within the family, have not yet been reached. Therefore, the new possibilities offered by the study of the distribution of chemical characters might be very useful.

A typical property of the Compositae is the relatively rich occurrence of different types of the so-called "secondary metabolites", especially of essential oils and other terpenoids, acetylenic compounds, phytomelanes, caffeic acid and its esters, flavones and flavonols, and, to a more limited extent, alkaloids. The possibilities of using these phytochemical characters for taxonomic studies have been recently thoroughly reviewed by Hegnauer (1964). It is typical of the recent rapid progress in the field of phytochemistry, that much new material has since been accumulated and that extension of Hegnauer's conclusions already seems desirable.

In this article more recent facts are presented regarding the occurrence in the Compositae of those sesquiterpenes which are regarded as typical members of their class. Sesquiterpene lactones are undoubtedly such a group of compounds (e.g. II and III), and the number of lactones of this type now isolated exceeds 170. The history of sesquiterpenes is very old, the compounds santonin or lactucin (VII) having been known for well over a hundred years. However, it is only during the last 15 years that the knowledge of this group of compounds has developed to its present state. One of the reasons for this development is the property of certain sesquiterpene lactones to give the easily recognizable compound chamazulene. This property was discovered by our school at the beginning of fifties. Other properties of sesquiterpene lactones, such as their bitterness and toxicity represented another stimulus to their wider study. Sesquiterpene lactones represent, for example, the typical bitter principles in Compositae (lactucin, lactucopikrin) as well as providing the toxic principles geigerin, helenalin and others. At present, these substances are systematically being studied in several laboratories; in addition to our own team, the schools of Herz, Romo, and Geissman are also active. This is evident from several reviews devoted to this theme (Herout, 1966a,b; Steelink and Spitzer, 1966; Herz, 1968).

However, sesquiterpene lactones cannot be considered as the only terpenoids which are useful in taxonomy. The study of the components of the Senecioneae tribe led to the discovery of compounds with an eremophilane skeleton (e.g. XIX) in many species of this group. As can be seen from the study of Novotný and Šorm (1965), furans derived from substances having an eremophilane basis characterize the Senecioneae. Their biogenetic relationship with the derived sesquiterpene lactones will be discussed later.

II. The Definition of Sesquiterpene Lactones as Taxonomic Characters

Sesquiterpene lactones occur to a small extent outside the Compositae and so it is necessary to describe the typical features of sesquiterpenoids of this family. These are mainly γ-lactones, where the lactone ring is usually formed by oxidation of one methyl group of the isopropenyl group in a hypothetical precursor (I) to a carboxyl group. This carboxyl group is then lactonized with the hydroxyl group formed by the oxidation of one of the possible α-methylene groups adjacent to the isopropenyl group (see Fig. 1). The presence of specific enzyme systems catalysing only this type of oxidation is evidently necessary, although it should be pointed out that we have no actual knowledge regarding either the character or the specificity of enzymes involved. Lactones of types different from those shown in Ib, for example where the γ-lactone ring is formed by oxidation of methyl groups at some other position (II and III), are not typical of the Compositae. Recently, γ-lactones of the aristolochia lactone type (II) have also been found in the Compositae. However, in all cases they are dilactones, where the formation of the second lactone ring evidently takes place as a secondary process, while the first lactone group has the typical

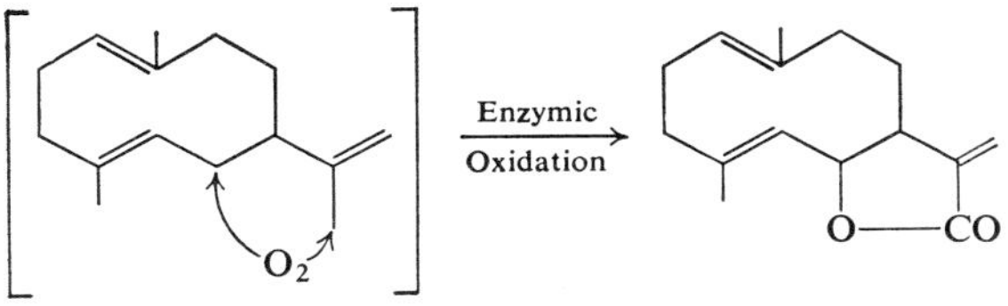

(Ia)   (Ib)

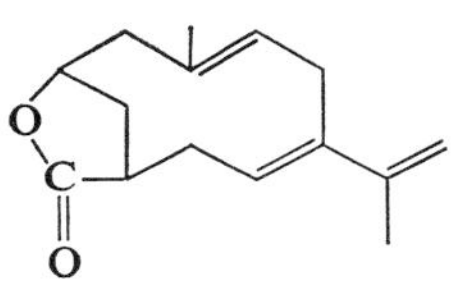

(II) Aristolochia lactone 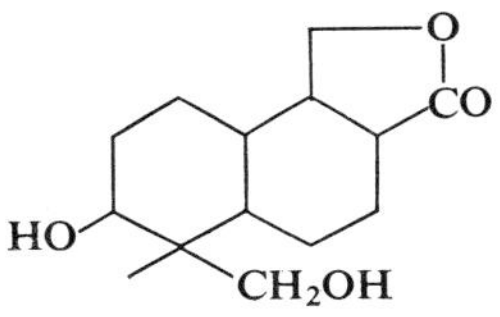   (III) Iresin 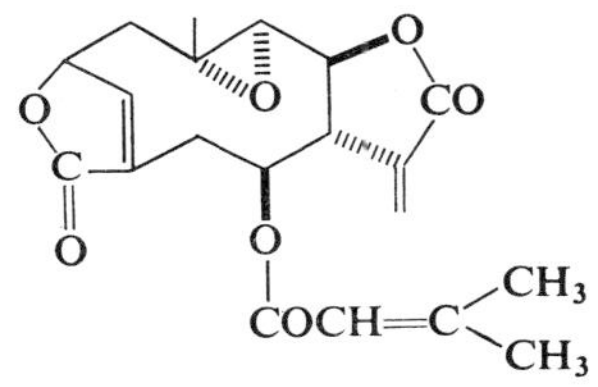   (IV) Elephantin

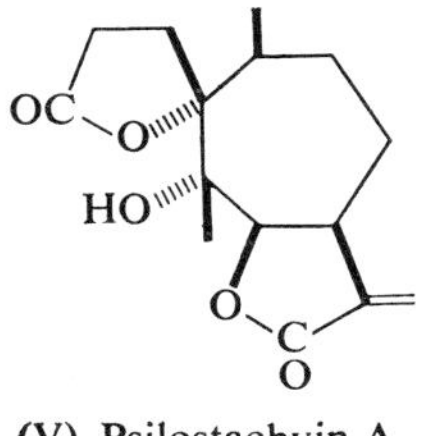

(V) Psilostachyin A 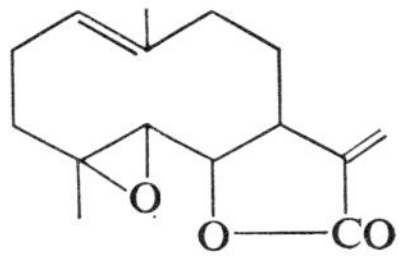   (VI) Parthenolide

Fig. 1

structure (Ib); see, for example, elephantin (IV) (Kupchan et al. 1966) or psilostachyin A (V) (Miller et al. 1965). It should also be noted that typical Compositae lactones have been found in other taxa. For example, parthenolide (VI), isolated originally from *Chrysanthemum parthenium* L. (Souček et al. 1961), was recently found in two representatives of the Magnoliaceae, i.e. in *Magnolia grandiflora* (Panizzi, 1965) and *Michelia champaca* (Govindachari et al. 1965). Magnoliaceae are a much more primitive family than the Compositae. The detection of identical metabolites in such phylogenetically distant families, does not, however, diminish the importance of the sesquiterpenes as taxonomic guides, since similar parallelisms are known for other groups of compounds as well as for morphological characters. In all cases, it is the whole gamut of characters that are important and usually enables differentiation between relationship and parallelism.

III. THE PROBABLE BIOSYNTHETIC PATHWAY TO SESQUITERPENE LACTONES

The known sesquiterpene lactones of the Compositae can be divided into five types, according to their carbon skeleton. Their probable biogenetic relationship is indicated schematically in Fig. 2, where only the essential skeletons of the compounds are given. It is not necessary to repeat in detail the mode of formation of sesquiterpenoids from farnesyl pyrophosphate since these studies have already been adequately summarized (e.g. Hendrickson, 1959). It is not known, however, whether the oxidation of primary precursors in the formation of the γ-lactonic grouping takes place simultaneously with the actual formation of the sesquiterpenoid skeleton, or whether the two processes are consecutive. It seems evident that the indicated biosynthetic pathway diverges at the stage of germacranolides (cf. Ognjanov et al. 1958; Barton and de Mayo, 1957), one path leading to eudesmanolides (also called santanolides) and by subsequent methyl migration of $C_{(14)}$ to eremophilanolides, and the other leading to guaianolides which, by a formally similar methyl migration, give pseudoguaianolides [a name originally used by Herz but another name, ambrosanolides, has been more recently proposed by Šorm and Dolejš (1965)]. A hypothesis has been put forward (Steelink and Spitzer, 1966) by which both types could be formed from a common precursor.

These five structures can be considered as representing the major types of sesquiterpene lactone which occur relatively frequently in the Compositae.The structures which differ only by alternative closing of the lactone ring (Fig. 2) are both well known to occur, with the exception of eremophilanolides. There are, of course, a series of lactones which are related to each of these main types, usually being formed by simple chemical changes. Some of these changes were observed to take place during the isolation of the lactones (i.e. not *in vivo*) and the compounds formed thus have no practical value in taxonomy. Such is the case in the formation of both saussurealactone and temisin by Cope reaction from precursors of the germacranolide type (Rao *et al.* 1961), or lumisantonin by a photoreaction (Barton and Gilham, 1959). However, other transformations probably take place in the plant. In such cases they are related to the gene-enzyme complex in the plant and their occurrence is of value for systematic classification. As an example of similar transformations of sesquiterpene lactones, the guaianolide precursors in certain species of *Xanthium* or *Iva* give rise to lactones with an open five-membered ring, (i.e. to xanthinin, Deuel and Geissman, 1957), xanthumin (Minato and Horibe, 1965), and ivalbin (Herz *et al.* 1967) (Fig. 2). Carabrone (Fig. 2) from *Carpesium abrotanoides* L. (Minato *et al.* 1964) is characterized by a similar structure, containing an

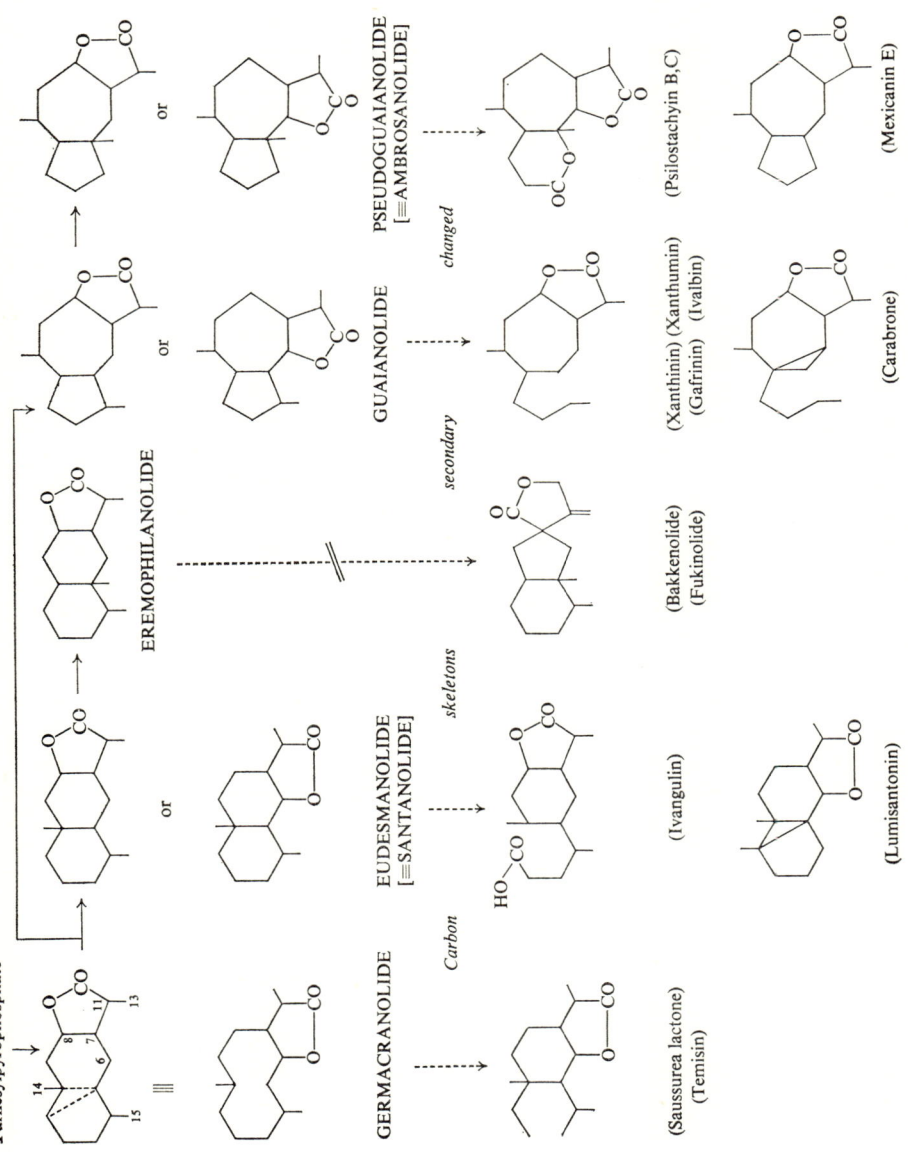

FIG. 2. Biogenetic relationship of the sesquiterpene lactones of the Compositae.

additional three-membered ring. Similarly, the formation of psilostachyin A, B and C (Fig. 2) in *Ambrosia psilostachyia* (Miller *et al*. 1965; Mabry *et al*. 1966; Kagan *et al*. 1966) could take place by the opening of the five-membered carbocyclic ring of the pseudoguaianolide precursor and, similarly, the formation of vermeerin which occurs in *Geigeria africana* (Anderson *et al*. 1967) by a slightly different opening of the same ring. The formation of the nor-sesquiterpene lactone mexicanin E in *Helenium mexicanum* can occur by the loss of a C_{14}-methyl in the course of the rearrangement of the guaianolide precursor to a substance of pseudoguaianolide type (cf. Romo *et al*. 1963). The bakkenolides or fukinolides (Abe *et al*. 1967, 1968; Naya *et al*. 1967), described recently, probably originate from an eremophilane precursor and show that the biosynthetic pathway in certain plants of the Senecioneae tribe can be further extended.

The great variability in structure of the presently known lactones in different plants consists not only of variations in the carbon skeleton, but also in subsequent changes in the sesquiterpene precursors, caused primarily by oxidation, giving rise to the formation of hydroxyl groups, carbonyl groups, oxide bridges and, less frequently, to additional carboxyls or lactonic functional groups. Very often the hydroxyls formed are then esterified. The variability in the esterifying acids (i.e. acetic acid or four- or five-carbon unsaturated or hydroxylated aliphatic acids) further increases the numbers of different sesquiterpenoids of this type which have been isolated so far. The extent of oxidation is often very specific of single tribes, subtribes or even genera of the Compositae.

Sesquiterpene lactones with carbon skeletons which do not fit into the scheme shown in Fig. 2 have yet to be obtained from composites. Hence, although substances with a bisabolane or cadinane skeleton and other sesquiterpene substances of different structures are known to occur in the Compositae (see Hegnauer, 1964), their biosynthesis is probably brought about by a different system of enzymes than that which is involved in the synthesis of sesquiterpene lactones.

IV. SESQUITERPENE LACTONES AND THE CLASSIFICATION OF THE COMPOSITAE INTO TRIBES

The relationship between the biogenetic types of sesquiterpene lactones found in the Compositae and the classification of this family is shown in Table I. The thirteen tribes into which Compositae are usually subdivided (see e.g. Hoffmann, 1894) are shown; the sub-division into subfamilies Asteroideae and Cichorioideae is not followed, and the Ambrosiaceae, considered first by Cassini as an independent family, is included provisionally in the Heliantheae.

The second column in Table I shows the number of species from which sesquiterpene lactones have been isolated, the structure of which (or at least the carbon skeleton) is reliably known. Reports of the occurrence of sesquiterpene lactones which contain insufficient data as to their structure are omitted

TABLE I

Frequency of occurrence of sesquiterpene lactones in the family Compositae

Tribe	Number of plant species investigated	Number of lactones of different types per tribe[a]					
		Germacra-nolides	Santa-nolides	Eremophila-nolides	Guaia-nolides	Ambrosa-nolides	C-Skeleton Changed
Vernoniae	2	3	—	—	—	—	—
Eupatoriae	4	2	—	—	2	1	—
Astereae	—	—	—	—	—	—	—
Inuleae	8	3 (4)	8 (11)	—	5	1	3 (4)
Heliantheae	31	1	5 (9)	—	2 (4)	16 (28)	10 (14)
Helenieae	32	12	1	—	1	34 (61)	3 (4)
Anthemidae	45	—	13 (39)	—	14 (23)	—	1
Senecioneae	4	—	—	6	—	—	3 (4)
Calendulae	—	—	—	—	—	—	—
Arctotidae	—	—	—	—	—	—	—
Cynareae	20	10 (19)	—	—	6 (7)	—	—
Mutisieae	—	—	—	—	—	—	—
Cichorieae	9	—	1	—	3 (12)	—	—

[a] Figures in brackets refer to the numbers of species per tribe which contain a particular lactone type.

from consideration. Because of the rapid development of knowledge in this area, the earlier reviews (Hegnauer, 1964; Šorm and Dolejš, 1965; Steelink and Spitzer, 1966; Merkel, 1966) are all considerably outdated. The data presented here (170 compounds from 155 plants) have been compiled mainly on the basis of the literature up to the end of 1967.

The compounds shown in subsequent columns of Table I are listed according to which of the five basic types they belong to; the last column comprises all the sesquiterpene lactones whose skeleton has undergone a subsequent change (with the exclusion of those compounds where the change evidently took place during isolation). The figures express the number of different compounds of a given type found in the corresponding tribe. Certain sesquiterpene lactones are found in several plants; for example α-santonin occurs in more than twenty *Artemisia* species (Anthemidae), tenulin in eleven and helenalin in seven *Helenium* spp. etc.; the sum of all these repeated occurrences of sesquiterpene lactones in a tribe is given in brackets.

A number of conclusions can be drawn from the data in Table I. First, one can see that lactones have not yet been found in all tribes; precise data is lacking regarding their presence in the Astereae, Calendulae, Arctotidae and Mutisieae. This does not mean, however, that lactones are absent from these tribes. For example, a sesquiterpene lactone was isolated from *Calendula officinalis* L. (Suchý and Herout, 1961) whose structure or skeleton has not yet been determined. In *Conyza dioscoridis* Desf. (Astereae), Saleh (1957) found a non-glycosidic bitter-tasting substance, which was presumably a sesquiterpene lactone. However, it is possible to state—in agreement with certain analyses carried out in our laboratories—that substances of this type are not of frequent occurrence in the four tribes mentioned.

Some of the tribes are characterized by the presence of only one type of sesquiterpene lactone and thus use only a limited part of the overall bio-synthetic pathways (Fig. 2). Vernoniae, for example, (in the two cases studied) stop at the germacranolides; in view of the small number of plants studied this fact has only limited value at present. The case of Cynareae is much more significant. Most species of this tribe cannot bring about further cyclization to the germacranolide skeleton, although in certain genera (*Cynara* and *Saussurea*) the ability to produce guaianolides is well pronounced. It is worth mentioning that formation of the eudesmane skeleton in species of this tribe has not yet been described.

Many of the tribes contain several different types of lactone. The Anthemidae, one of the richest in lactones, is typical in this respect. In addition to germa-cranolides, the lactones of this tribe are almost equally formed by guaianolides and santanolides. However this tribe, which is mainly distributed in the Old World (Europe–Africa–Asia, especially the Mediterranean area), does not produce a single compound of either the pseudoguaianolide or changed C-skeleton type. The only exception is the presence of lumisantonin which was reported in *Artemisia kurramensis* Quazilbash (Satoda *et al.* 1959); it is probable, however, that in this case the compound is produced by a photoreaction and hence an artifact. The Cichorieae, which is usually considered to be an independent subfamily (Liguliflorae), is characterized by the formation of guaianolides, especially lactucin and lactucopicrin (Fig. 3; VII, VIII; cf., e.g., Holzer and Zinke, 1953; Barton and Narayanan, 1958; Dolejš et al. 1958). The occurrence of compounds of the santanolide type—tuberiferin (IX) in *Sonchus tuberifer* Svent. (Bairerra et al. 1967)—is as yet untypical.

Two of the tribes of the Compositae are distinctly different from those mentioned above, in that they are restricted either exclusively (Helenieae) or predominantly (Heliantheae) to the American continent. The Helenieae

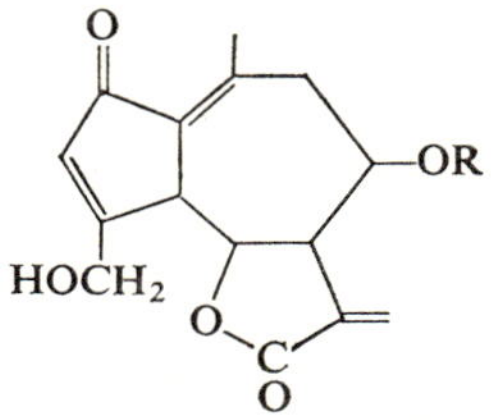

(VII) R = H.   Lactucin

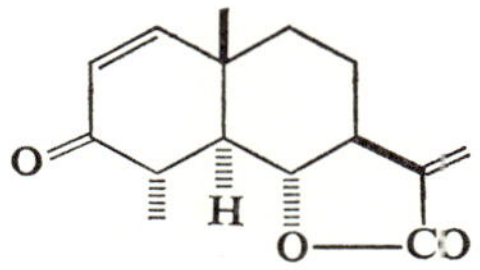

(IX) Tuberiferin

(VIII) R = COCH₂—⟨benzene⟩—OH. Lactucopicrin

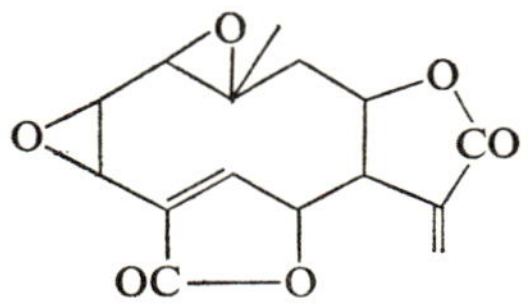

(X) Mikanolide 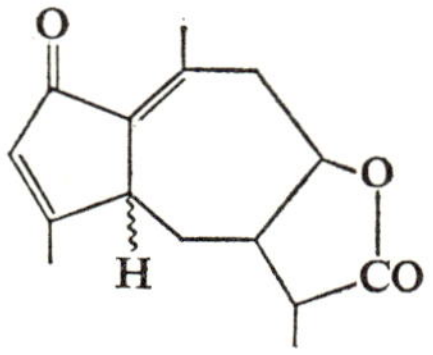   (XI) Virginolide 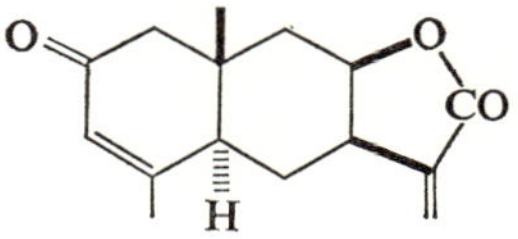   (XII) Pinnatifidine

FIG. 3

contain numerous lactones of pseudoguaianolide (ambrosanolide) type, and the occurrence of substances earlier on the biosynthetic pathway, i.e. mikanolide (X) from *Gaillardia fastigiata* Greene (Herz et al. 1966b) and virginolide (XI) from *Helenium virginicum* Blake (Herz and Santhanam, 1967) are exceptional. The only santanolide isolated from this tribe is pinnatifidine (XII) from *Helenium pinnatifidum* (Nutt.) Rybd. (Herz et al. 1962).

The Heliantheae can be best considered if those plants which belong to the evolutionary young family of Ambrosiaceae are separated from it (as has already been proposed by a number of taxonomists for other reasons). The Ambrosiaceae contain the genera *Iva*, *Ambrosia*, *Franseria* and *Xanthium*. The results of doing this are shown in Table II. It can be seen that the Heliantheae proper are very similar to the Helenieae, the occurrence of ambrosanolides

TABLE II

Frequency of occurrence of lactones in the tribe Heliantheae

	Number of plants investigated	Germacra- nolides	Santa- nolides	Guaia- nolides	Ambrosa- nolides	Carbon skeleton changed
Tribe:						
Heliantheae	6	1	—	—	5 (6)	2
Family:						
Ambrosiaceae	25	2 (3)	5 (9)	2 (4)	11 (22)	8 (12)

(pseudoguaianolides) again being predominant. Substances with altered skeletons are represented only by zaluzanin A and B (Fig. 4, XIIIa, b) from *Zaluzania augusta* (Lag.) Schultz (Romo *et al.* 1967), both having an extra three-membered ring in the guaianolide skeleton. The Ambrosiaceae is characterized by the frequent occurrence of lactones of all types except the eremophilanes. Ambrosanolides are the most numerous and the fact that this family contains a large number of lactones which are biosynthetically "advanced" should not be overlooked.

It is less easy to express an opinion on the position of two tribes, Eupatoriae and Inuleae. In neither case has a sufficient number of species been studied. Each group shows the presence of one lactone of the ambrosanolide type: stevin (XIV) from *Stevia rhombifolia* H.B.K. (Ríos *et al.* 1967) of the Eupatoriae, and geigerinin (XV) from *Geigeria aspera* Harv. (Inuleae) (Villiers and

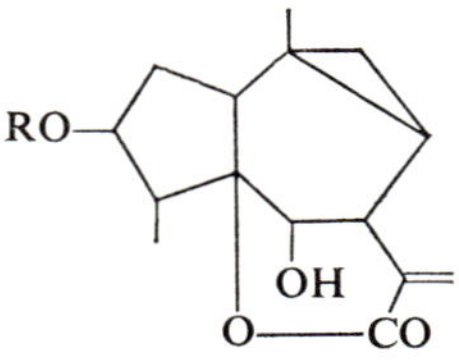

(XIIIa) R = H. Zaluzanin A
(XIIIb) R = Ac. Zaluzanin B

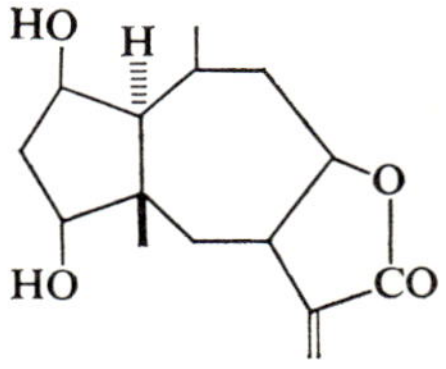

(XIV) Stevin

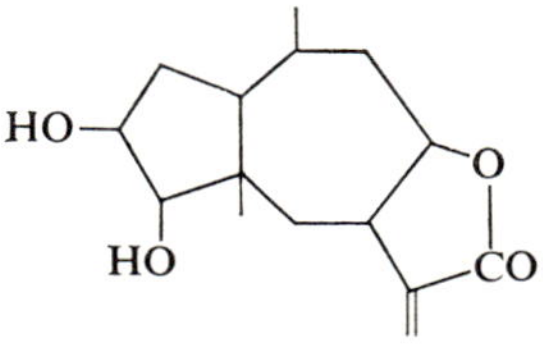

(XV) Geigerinin

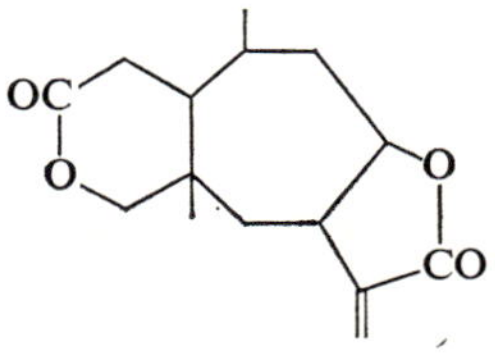

(XVI) Vermeerin

Fig. 4

Pachler, 1963). *G. aspera* and *G. africana* Gries., also contain a substance with a more extensively changed skeleton of the pseudoguaianolide type: vermeerin (XVI) (Anderson *et al.* 1967). Thus, *Geigeria*, on the basis of its chemistry, shows that a certain relationship exists between the Inuleae and the evolutionary very ancient tribe of Heliantheae.

On the basis of a similar evaluation of all thirteen Compositae tribes, the Senecioneae remains completely apart; it is the only one to contain the eremophilanolide type lactones (Table I). Significantly, this tribe is already known to be unusual in that senecio alkaloids are widespread in its species. The rare eremophilanolide lactones are presumably formed by aerial oxidation of the much more widespread furoderivatives (XVII) (see Novotný and Šorm, 1965; Hikino *et al.* 1962). It is not easy to decide in all instances whether this auto-

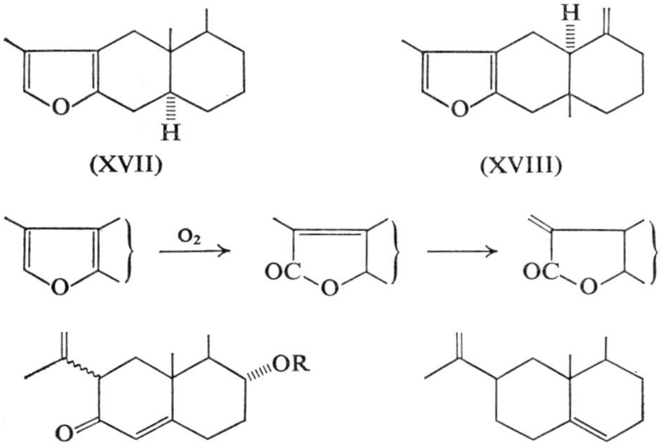

FIG. 5. Eremophilane type lactones.

oxidation (see Fig. 5), observed originally in the case of atractylone (XVIII), takes place within the plant or after the isolation of the compound.

In addition to the relatively highly oxidized derivatives of an eremophilane type, lower oxidation products were also often observed in species of the Senecioneae, as for example compounds of the petasin type (XIX) (Aebi *et al.* 1958) or even the hydrocarbon eremophilene (XX) (Křepinský *et al.* 1968). A complete list of compounds of the eremophilane type found in this tribe is shown in Table III.

From the data presented here, one can conclude that the division of the Compositae, determined originally entirely on the basis of morphological observations, is quite justified when the distribution of such widely divergent compounds as the sesquiterpenoids, especially sesquiterpene lactones, is taken into account. The proposed biosynthetic relationships of the compounds present in the composites (Fig. 2) makes it possible to express tentatively a dynamic

6

TABLE III

The occurrence of eremophilane derivatives in the Senecioneae

Plant	Less-oxygenated derivatives	Furanoderivatives	Lactones	References
Adenostyles alliariae (Gouan.) Kern	eremophilene	adenostylone isoadenostylone neoadenostylone	unidentified	Harmatha *et al.* 1968
Cacalia decomposita A. Gray	—	decompostin	—	Rodríguez-Hahn *et al.* 1968
Euryops floribundus N.E. Br.	—	euryopsol[a] euryopsonol[a]	—	Rivett and Woolard, 1967
Homogyne alpina (L.) Cass.	—	—	bakkenolide A[b]	Harmatha *et al.* 1968
Ligularia fisheri (Ledeb.) Turcz.	—	ligularol (= petasalbin) ligularone	—	Ishii *et al.* 1965
	—	furanoligularenone	—	Patil *et al.* 1965
Petasites albus (L.) Gaertn.	eremophilene	furoeremophilane petasalbin albopetasin angelyljaponicin albopetasol	6-hydroxy-eremophile-nolide	Novotný *et al.* 1962 Hochmannová *et al.* 1962
Petasites hybridus (L.) Gaertn. Meyer et Scherb.	eremophilene	—	—	Hochmannová and Herout, 1964
	petasin, isopetasin *S*-petasin, *S*-isopetasin	—	—	Stoll *et al.* 1956 Aebi *et al.* 1958
	—	furoeremophilane 9-hydroxyfuroeremo-philane	eremophilenolide petasolide A, B *S*-petasolide A, B	Novotný *et al.* 1962, 1963, (Novotný and Herout) 1966, 1968

Species		9-ketofuroeremophilane furanopetasin (dimethoxydihydrofuroeremophilone)			References
Petasites japonicus (Sieb. et Zucc.) Maxim. subsp. *giganteus* Kitam.	eremophilene	—		—	Novotný and Herout, 1965
	petasitin	—		—	Naya and Tagaki, 1968
	—		furoeremophilone petasalbin albopetasin angelyljaponicin	6-hydroxyeremophilenolide	Novotný *et al.* 1965
	—			bakkenolide A, B, C, D[b]	Abe *et al.* 1967, 1968
	—			fukinolide[b] S-fukinolide[b]	Naya *et al.* 1967
Petasites kablikianus Tausch. ex Bercht.	eremophilene		furoeremophilane kablicin angelyljaponicin diangelyljaponicin kablikopetasin	6-hydroxyeremophilenolide	Novotný *et al.* 1968
Petasites paradoxus (Retz.) Baum.	—		furoeremophilane kablicin angelyljaponicin diangelyljaponicin kablikopetasin	6-hydroxyeremophilenolide	Novotný *et al.* 1968
Petasites spurius (Retz.) Reichenb.	eremophilene		petasalbin albopetasin	—	Novotný and Herout, 1962

[a] Isolated after saponification only.
[b] Contains a changed eremophilane skeleton.

TABLE IV

The structures and the distribution of germacranolides in the Cynareae

Lactone	Formula	Isolated from plants	References
Albicolide	(structure: CH₂OH; HOCH₂; 1,4,5,6,7,8,10,11,12,14,15; CO, O)	*Jurinea albicaulis* Bge. var. *kilaea* (Aznav.) Stoi. et Stef.	Suchý et al. 1967b
Arctiopicrin	(structure: OCOCH·CH₃·CH₂OH; HOCH₂; CO, O)	*Arctium lappa* L. *A. minus* (Hill.) Bernh. *A. nemorosum* Léj. et Court. *A. tomentosum* Hill.	Drozdz, 1967 Cavalitto et al. 1945 Drozdz, 1967 Drozdz, 1967
Cnicin	(structure: COC=CHCH₂OH·CH₂OH; HOCH₂; CO, O)	*Centaurea calcitrapa* L., *C. diffusa* Lam., *C. iberica* Trev., *C. micranthos* I.F. Gmel., *C. ovina* Pall. *C. stoebe* (L.) Sch. et Thell. *Cnicus benedictus* L.	Drozdz, 1967 Suchý and Herout, 1962 Korte and Bachmann 1958
Costunolide	(structure: CO, O)	*Saussurea lappa* C. B. Clarke	Rao et al. 1958

Compound	Structure	Species	Reference
Jurineolide	CH_2OH; $OCOC(CH_3)=CH\cdot CH_2OH$; CO; $HOCH_2$; O	*Jurinea cyanoides* (L.) Rchb.	Suchý *et al.* 1968b
Onopordopicrin	$OCOCCH_2OH$ ($=CH_2$); CO; $HOCH_2$; O	*Onopordon acanthium* L.	Drozdz, 1967
Salonitenolide	CO; O; $HOCH_2$; OH	*Centaurea salonitana* Vis.	Suchý *et al.* 1967a
Salonitolide	CO; O; $HOCH_2$; OH	*Centaurea salonitana* Vis.	Suchý *et al.* 1965
Scabiolide	CO; O; OH; $COC(CH_3)\cdot CH_2OH$; $AcOCH_2$	*Centaurea scabiosa* L.	Suchý *et al.* 1962

view of the phylogenetic relationships between the tribes, and to compare this view with conclusions based entirely on non-chemical characters (cf. Cronquist, 1955).

If one includes the Ambrosiaceae, the Heliantheae, which are usually considered the most primitive and hence the oldest taxon, in fact represents the chemically most developed group, if the multiplicity of sesquiterpene lactone types is taken into account. They are able to synthesize all types of substances of this series. The Vernoniae and the Cynareae, on the contrary are only able to synthesize a limited number of these compounds, which points to their relatively low degree of evolution. It is interesting that the views of certain taxonomists agree with this opinion (Leonhardt, 1949; Augier and Du Merac, 1951).

The advanced position of the Helenieae, connected quite evidently with the evolution of Heliantheae, should not be overlooked. This is clearly seen by comparison with the Anthemideae which cannot synthesize compounds more "advanced" than the santanolides in the first branch of the biosynthetic pathway, or in the case of the second branch, than the guaianolides. The known geographic differences support the view of a separate evolution of the Heliantheae and Heleniae on the one hand and the Anthemidae on the other. Finally it is difficult to disregard the independent position of the Senecioneae, a tribe characterized by the exclusive occurrence of substances of the eremophilane type.

V. Sesquiterpene Lactones and the Classification within Tribes

A. cynareae

Nine germacranolides (see Table IV; structures are from Suchý *et al.* 1968a), which are structurally strikingly uniform, have been isolated from the Cynareae. The uniform position of the double bonds is typical; most contain three double bonds, situated between carbon atoms 4 and 5, 10 and 1 and 11 and 12. The γ-lactone ring is closed predominantly at C-6 or in a few cases at C-8. In the majority of cases the second of the two α-methylene groups vicinal to the isopropyl is also oxidized; costunolide, representing the basic unoxidized lactone, and albicolide are the only exceptions. The Cynareae is characterized by the fact that all lactones having a primary alcoholic group formed by the oxidation of the methyl group at C-15. In both substances isolated from *Jurinea* species the methyls at C-14 and C-15 on the ten-membered ring are both oxidized. It will be interesting to see whether further investigation of this genus shows this to be a regular feature.

In addition to the uniformity in the structure of this group of lactones, it is worth noting that their stereochemistry is also similar, which indicates they are formed by a uniform biosynthetic pathway. The configuration on carbons 6, 7 and 8, is similar to that found in lactones of the santanolide group, for example in artemisin (Fig. 6, XXI; Sumi, 1958).

As mentioned earlier some plants of this tribe also contain guaianolides;

for example *Cynara scolymus* L. and *C. cardunculus* L. contain cynaropicrin (XXII; Suchý *et al.* 1960), *Saussurea pulchella* Fisch. contains saurin (XXIII) (Kushnir and Kuzovkov, 1966) and *S. lappa* C. B. Clarke ("Costus root")

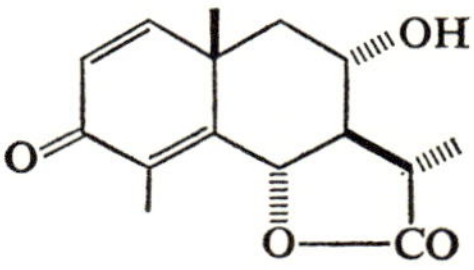

(XXI) Artemisin

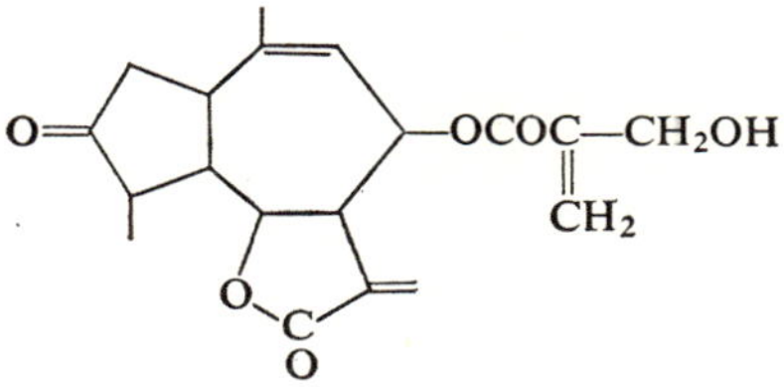

(XXII) Cynaropicrin

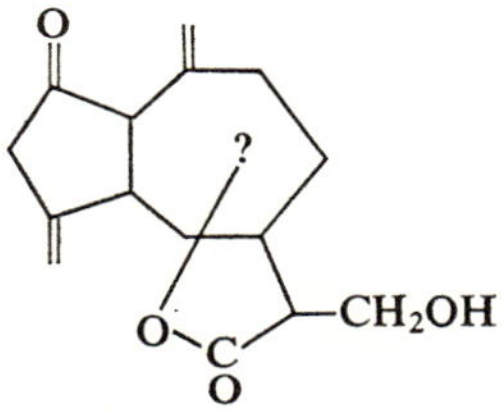

(XXIII) Saurin

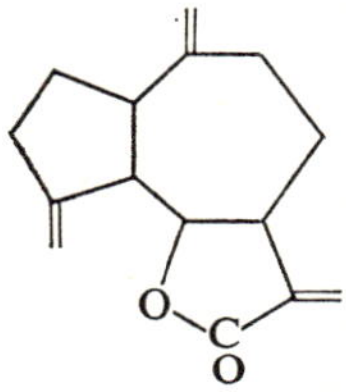

(XXIV) Dehydrocostus lactone

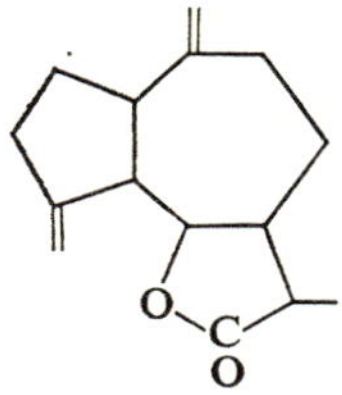

(XXV) Costuslactone

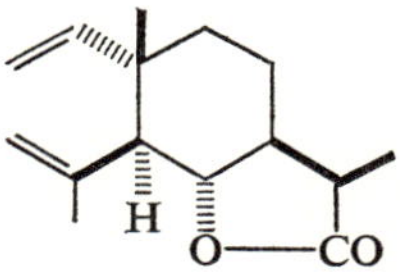

(XXVI) Saussurea lactone

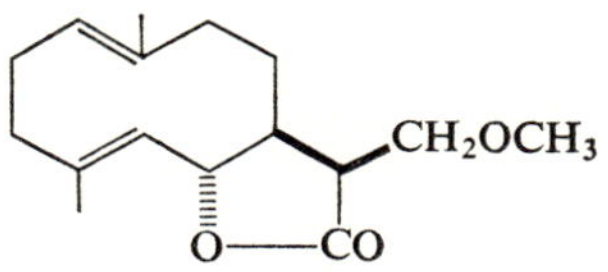

(XXVII) Methoxydihydrocostunolide

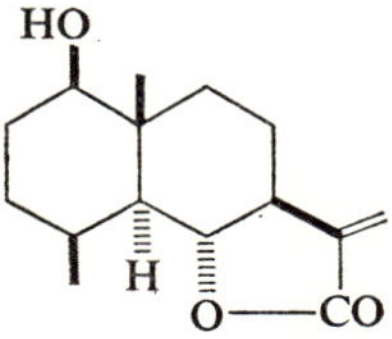

(XXVIII) Balchanin

FIG. 6.

contains a whole group of lactones, differing only in the number of double bonds. The most important of these are dehydrocostuslactone (XXIV) (Romaňuk *et al.* 1958), its dihydroderivative, costuslactone (XXV) (Kulkarni *et al.* 1963), and dihydrocostuslactone (cf. Naves, 1949). These substances have been isolated both from the essential oil and from the extract (in addition to other compounds probably formed by a change similar to the case of saussure-alactone (XXVI), or by addition of methanol-methoxydihydrocostunolide (XXVII) (Kulkarni *et al.* 1961) together with germacranolides) and it is interesting to note that the basic lactone of this plant—costunolide—is extremely easily transformed by acid catalysed reaction to derivatives of the santanolide type. However, no derivative of the latter type has been found in costus, which illustrates that the cyclization of costunolide or some other precursor to derivatives of the guaianolide type must be under enzymic control. In comparison, it is worth mentioning that in another plant rich in costunolide and other germacranolides—*Artemisia balchanorum* H. Krasch (Anthemideae)—a santanolide derivative was also found, i.e. balchanin (XXVIII) (Suchý, 1962). This supports the hypothesis, expressed earlier, on the difference in the enzymic systems in Cynareae and Anthemideae.

B. ANTHEMIDEAE

It seems premature to draw conclusions regarding the systematics of single subsections of this tribe, although the amount of data available is appreciable. However, one should mention the situation pertaining to the genus *Artemisia* (Geissman, 1966). This genus consists of some 200-odd species and is commonly divided into four main sections, primarily on the basis of floral morphology. No sesquiterpene lactones have been found so far in the Dracunculus section. Santonin is prevalent only in a limited number of the members of the Seriphidium section which contains other santanolides and several examples of this structural type are found in members of the Abrotanum section. Guaianolides are also found in both *Seriphidium* and *Abrotanum* species, as well as in those of the Absinthium section. It is obvious, therefore, that more information is needed before the chemotaxonomic value of the sesquiterpenes of the *Artemisia* can be assessed.

C. HELIANTHEAE AND HELENIEAE

Both the related tribes, Heliantheae and Helenieae contain a plethora of sesquiterpene lactones, especially those of the pseudoguaianolide type. A large number of them, however, occur only in one individual species, and very few occur sufficiently frequently to permit conclusions to be drawn regarding the relationship of the plants from which they were isolated. In Table V, the distribution of several such pseudoguaianolides in both tribes is shown. For reasons mentioned earlier, the subtribe Ambrosiineae (identical with the independent family Ambrosiaceae) has been separated from the Heliantheae tribe. The results are based mainly on the systematic studies of Herz (for the genus *Iva* cf. Herz *et al.* 1967c and previous papers, for the genus *Helenium* cf. Herz and Lakshmikantham 1965 and preceding papers, for the genus *Gaillardia* see Herz *et al.* 1967b and preceding papers) and other authors (e.g., Geissman and Toribio, 1967). From this data, several simple conclusions can

TABLE V

The frequency of occurrence of some pseudoguaianolides

Pseudo-guaianolide	Substitution pattern[a]	Heliantheae	(Ambrosiineae)			Helenieae		
		Parthenium	*Ambrosia*	*Iva*	*Hymenoclea*	*Helenium*	*Balduina*[d]	*Gaillardia*
Spathulin	1,3-dihydroxy, 5,8-diacetoxy	–	–	–	–	–	–	4/5[b]
Helenalin	Δ^1,3-keto, 5-hydroxy	–	–	–	–	6	1	2
Tenulin	See below[c]	–	–	–	–	11	–	–
Ambrosin	Δ^1,3-keto	1	2	–	1	–	–	–
Damsin	3-keto	–	2	–	–	–	–	–
Parthenin	Δ^1,3-keto, 10-hydroxy	1	1	1	–	–	–	–
Coronopilin	3-keto, 10-hydroxy	1	2	2	1	–	–	–

[a] The numbering of the common moiety of all lactones mentioned is shown in formula A.

[b] In one case (*Gaillardia grandiflora*), the plant enumerated represents a hybrid.

[c] Structure of tenulin is shown by formula B.

[d] This genus is more usually classified as a member of the Heliantheae.

A

B

be drawn: the occurrence of spathulin is limited to *Gaillardia* and tenulin to *Helenium*. The occurrence of helenalin, present in *Balduina*, *Gaillardia* and *Helenium*, is typical of the plants of the Helenieae. The Heliantheae, on the other hand, contains none of these three compounds, but does contain several sesquiterpene lactones which are peculiar to it, and occur both in the Heliantheae proper and in the derived Ambrosiaceae. These are ambrosin, parthenin and notably coronopilin which has been found so far in five genera (*Parthenium*, *Ambrosia*, *Iva*, *Hymenoclea* and *Franseria dumosa* Gray, the latter not being included in the table; Geissman and Turley, 1964).

Special attention should be drawn to the simultaneous occurrence of the sesquiterpenoid ilicic acid (Fig. 7; formula XXIX) in *Ambrosia ilicifolia* (Gray) Payne (Herz *et al.* 1966a) and in *Hymenoclea salsola* T. et G. (Geissman and Toribio, 1967). Morphological studies show that both plants have a rather close phylogenetical relationship. Thus, the presence of a common metabolite can be considered as significant. It points to a common, preserved enzyme system leading in both species to a derivative of the eudesmane type, which is

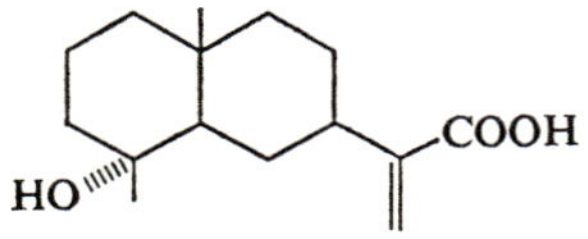

(XXIX) Ilicic acid

FIG. 7.

more "primitive" from the biochemical point of view than the other pseudo-guaianolides present. In any case, oxidation leading only to an acid without the typical formation of a γ-lactone ring, is unusual in the Compositae, and hence rather interesting.

D. SENECIONEAE

This tribe, sharply distinguished chemically from other tribes of Compositae by the occurrence of the senecio alkaloids and by the absence of the otherwise widely distributed acetylenic derivatives and phytomelanins, is distinguished also by a very rich distribution of sesquiterpenoids of the eremophilane type. We have tried, therefore (Novotný *et al.* 1966b), to utilize this fact for determining the taxonomic relations of several European species of *Petasites*. We started from the fact that the eremophilane derivatives present in *P. hybridus* (L.) G.M. et Sch. are characterized by oxidation exclusively in position 9 [see Table III and Fig. 8 for the formulae of the main components, for example, furanopetasin (XXX), 9-hydroxyfuroeremophilene (XXXIa), 9-ketofuroerem-ophilene (XXXIb)]. In contrast to this, *P. albus* (L.) Gaertn. typically contains derivatives oxidized at position 6, as for example petasalbin and its angelylester (XXXIIa,b). This indicates again the existence of specific enzymes in the two

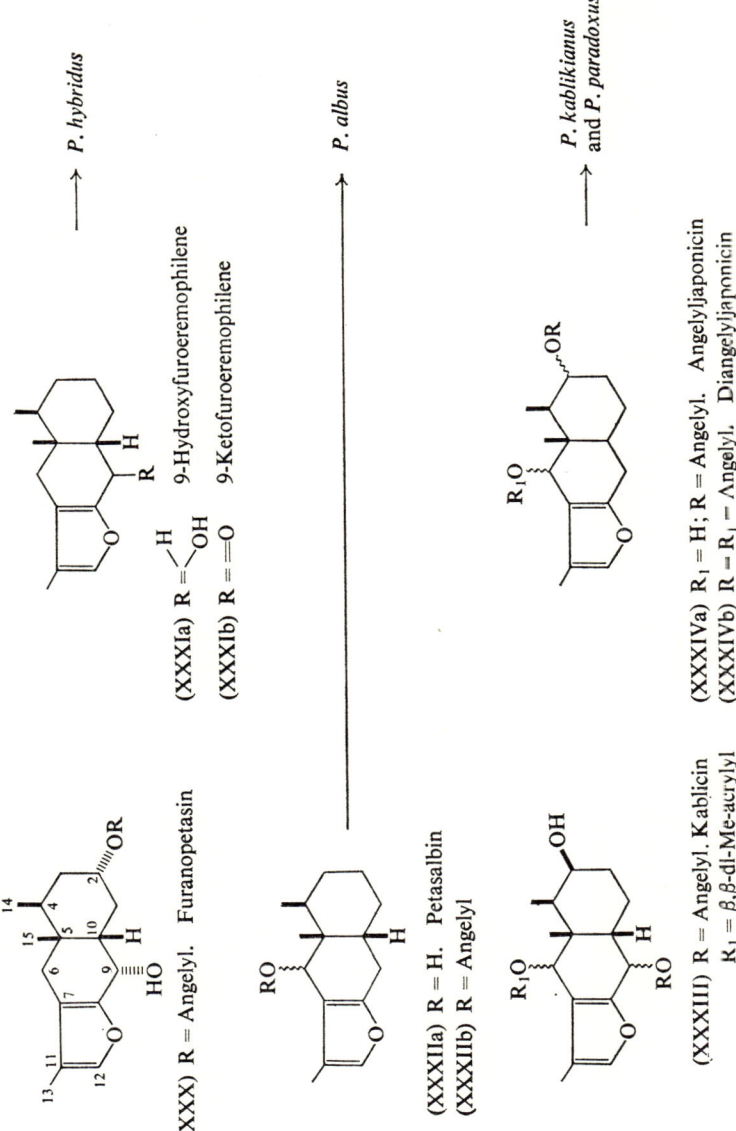

Fig. 8. Sesquiterpene lactones of *Petasites*.

species. In the natural hybrid of the two species, designated *P.* × *rechingeri* Hayek, metabolites of both parent species can be found, i.e. notably furano-petasin and petasalbin. This fact shows that a species occurring in the Giant Mountain and Carpathian area, i.e. *P. kablikianus* Tausch ex Bercht., cannot be merely a simple hybrid (*P. albus* × *P. hybridus*). On the basis of morpholo-gical studies, we concluded that *P. kablikianus* is an independent species of hybrid origin, which underwent a speciation process of its own. From the phytochemical point of view, it is interesting that the typical components of *P. kablikianus* are kablicin in which both position 9 and the position 6 (see formula XXXIII) and angelyl and diangelyljaponicin (XXXIVa,b) in which positions 6 are oxidized. We have also examined morphologically another European species, *P. paradoxus* (Retz.) Baum. characterized by an Alpine

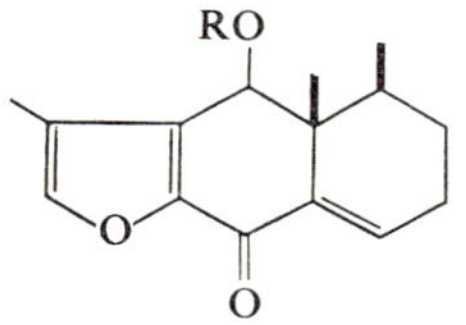

(XXXVa) R = Isobutyryl Adenostylone
(XXXVb) R = Angelyl Neoadenostylone

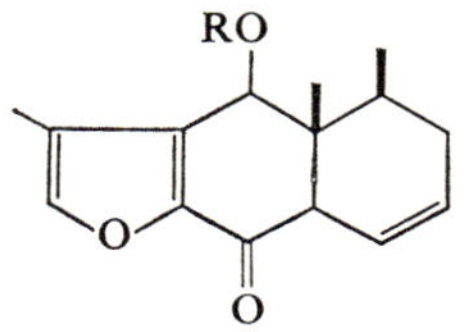

(XXXVI) R = Isobutyryl
Isoadenostylone

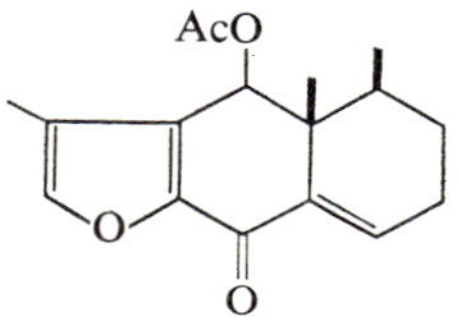

(XXXVII) Decompostin

Fig. 9. Sesquiterpene lactones of *Adenostyles* and *Cacalia*.

distribution, and we concluded that it is an Alpine parallel to *P. kablikianus*. A further phytochemical study (Novotný, *et al.* 1968b) has corroborated this view to the full, because the components of the eremophilane type in the two species are identical.

Adenostyles alliariae (Gouan.) Kern contains, according to the latest information (Harmatha *et al.* 1968b) a group of substances of eremophilane type; in addition to eremophilene, typical of the genus *Petasites*, it has adeno-stylone (see Fig. 9, XXXVa), isoadenostylone (XXXVI) and neoadenostylone (XXXVb). The presence of these compounds favours the classification of *Adenostyles* with the Senecioneae tribe and not with Eupatorieae, where it was classified earlier by certain authors. Mexican authors (Rodriguez-Hahn *et al.* 1968) have recently found decompostin (XXXVII) in *Cacalia decomposita* A. Gray. This compound only differs from adenostylone in its acyl group and shows a relation between *Adenostyles* and *Cacalia* from the phytochemical point of view; indeed, *Adenostyles* was separated from *Cacalia* by Cassini only in 1816.

On the basis of similar considerations, one can now examine the relationships between the genera *Homogyne* and *Petasites*. From *Homogyne alpina* (L.) Cass. a lactone $C_{15}H_{22}O_2$ was isolated (Harmatha *et al.* 1968a) identical with bakkenolide A (Fig. 10, XXXVIII) isolated also recently from *Petasites japonicus* Maxim. subsp. *giganteus* Kitam. (Abe *et al.* 1967, 1968). This lactone can be derived from an eremophilane precursor by a simple rearrangement (cf. Fig. 10) and this relation gives chemotaxonomic support to the relationship of the two genera mentioned. The same variety of *P. japonicus*, growing in Czechoslovakia in two localities, has afforded only petasalbin in its rhizomes. The Japanese authors, however, did not study the rhizomes but buds, and it is not without interest that a third group, studying the flower stems of the same

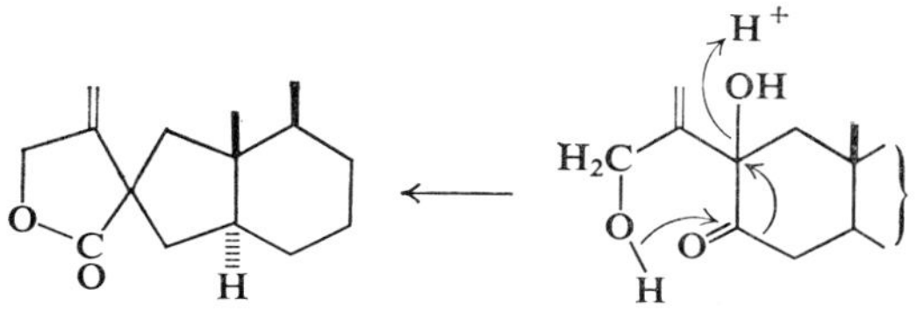

(XXXVIII) Bakkenolide A

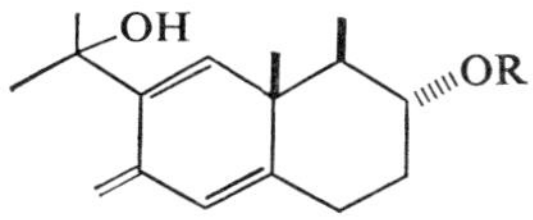

(XXXIX) R = Angelyl   Petasitin

FIG. 10. Sesquiterpene lactones of *Homogyne*.

plant (Naya and Tagaki, 1968) isolated an additional eremophilane derivative, analogous to petasin—i.e. petasitin (XXXIX). This indicates a certain danger in making conclusions from phytochemical studies based on examination of only one sort of tissue. However, notwithstanding this limitation the study of eremophilane derivatives from an as yet limited number of plants of the Senecioneae tribe shows very promising results.

VI. LIMITATIONS IN THE TAXONOMIC SIGNIFICANCE OF THE PHYTOCHEMICAL DATA

In view of the enormous number of presently known species of Compositae, the available data on the occurrence of sesquiterpene lactones and other sesquiterpenoids represent only a limited survey. In this respect, however, the much accelerated research of recent years, caused mainly by improvements in the isolation procedures and structure analysis, promises the more rapid addition of new data in the near future. One annoying feature is that negative results are only seldom published, although it is well known that the presence of sesquiterpenoids in Compositae is far from universal, even in such tribes as have been mentioned as being very rich in such constituents. This fact will appreciably limit the usefulness of phytochemical data, and represents an evident disadvantage in comparison with other methods—for example morphological, cytological or palynological studies although here too valid limitations apply; i.e. the comparison of components from morphologically differing parts, variability due to localities or to climatic differences and so on.

It is necessary also to examine any differences in the content of a given substance in individual plants. The existence of so-called chemical races (chemovars) is well-enough known and has already been studied by Stahl (1952a,b), who looked for the presence or absence of chamazulenogenic substances, for example, in *Achillea millefolium* L. We have made an interesting observation in the case of *Petasites hybridus*, which often propagates only vegetatively in any one locality; thus, the population from any other locality often displays conspicuous differences. While, for example, we found that plants from the majority of localities in Czechoslovakia contained furanoeremophilane derivatives, Swiss authors (Aebi *et al.* 1958) found in their plants exclusively the so-called petasin type of compounds. In a more systematic study we have examined (Novotný *et al.* 1966b) samples from more than fifteen different localities in France and Germany; individual populations were found to contain one or other of the above-mentioned chemovars in an approximate 1:1 ratio.

The most extensive study of this type is that of Miller *et al.* (1968) who studied the distribution of fifteen sesquiterpene lactones in sixty-two populations of *Ambrosia psilostachya* ranging from Canada to Mexico, and in four Mexican populations of the closely related *A. cumanensis* Kunth. Their results demonstrate that phytochemical differences within single populations are appreciable. In North America three lactones prevailed—coronopilin, parthenin and ambrosiol, while in populations from Galveston Island only the dilactones psilostachyin and psilostachyin B and C were present. Similarly, these latter lactones were also typical for *A. cumanensis*. The finding that within the *A. psilostachya* species, differentiation can be observed is remarkable; certain populations can only synthesize pseudoguaianolides while others are able to carry out a further enzymatic oxidation leading to pseudoguaianolides with a changed skeleton, i.e. to the dilactones mentioned. From the work by Miller *et al.* (1968) it can be concluded that the populations containing the dilactones are relatively younger, as the geological age of these offshore barrier islands is known not to exceed 5000 years.

VII. CONCLUSION

The views presented in this article are intentionally one-sided, since they concern only one character, the occurrence of sesquiterpene lactones or related sesquiterpenoids which have been thoroughly studied only very recently. They should, however, serve as a stimulus for further systematic studies. The partial conclusions made here may then be subjected to some criticism. Only the synthesis of all accessible phytochemical data on the plants of this complicated family can give a complete answer. It will show to what degree comparative phytochemistry can be used in the classification, and especially for conclusions concerning the phylogeny of the Compositae.

REFERENCES

Abe, A. (1968). Private communication.

Abe, A., Onoda, R., Shirata, K., Kato, T., Woods, M. C., and Kitahara, Y. (1967). Symp. Papers, "The Chemistry of Natural Products", p. 96, Kyoto.

Abe, A., Onoda, R., Shirata, K., Kato, T., Woods, M. C., and Kitahara, Y. (1968). *Tetrahedron Lett.* 369.

Aebi, A. Waaler, T., and Büchi, J. (1958). *Pharm. Weekbl. Med.* **93**, 397.

Anderson, L. A. P., de Kock, W. T., Pachler, K. G. R., and Brink, C. M. (1967). *Tetrahedron*, **23**, 4153.

Augier, J., and Du Merac, M. (1951). *Rev. Scient.* **3311**, 167.

Barreira, J. B., Bretón, J. R., Fajards, M., and Gonsáles, A. G. (1967). *Tetrahedron Lett.* 3475.

Barton, D. H. R., and Gilham, P. T. (1959). *Proc. chem. Soc.* **391**.

Barton, D. H. R., and de Mayo, P. (1957). *J. chem. Soc.* **150**.

Barton, D. H. R., and Narayanan, C. R. (1958). *J. chem. Soc.* **963**.

Cavallito, Ch. I., Bailey, I. H., and Kirchner, P. K. (1945). *J. chem. Soc.* **67**, 948.

Cronquist, A. (1955). *Am. Mid. Nat.* **53**, 478.

Deuel, P. G., and Geissman, T. A. (1957). *J. Am. chem. Soc.* **79**, 3778.

Dolejš, L., Souček, M., Horák, M., Herout, V., and Šorm, F. (1958). *Coll. Czech. Chem. Commun.* **23**, 2195.

Drozdz, B. (1967). Dissertation Thesis. University of Poznan.

Geissman, T. A. (1966). *J. org. Chem.* **31**, 2523.

Geissman, T. A., and Toribio, F. P. (1967). *Phytochemistry*, **6**, 1563.

Geissman, T. A., and Turley, R. J. (1964). *J. org. Chem.* **29**, 2553.

Govindachari, T. R., Joshi, B. S., and Kamat, V. (1965). *Tetrahedron*, **21**, 1509.

Harmatha, J., Novotný, L., Samek, Z., and Herout, V. (1968a). (In Press.)

Harmatha, J., Novotný, L., Samek Z., Herout, V. and Šorm, F. (1968b). *Tetrahedron Lett.* 1409.

Hegnauer, R. (1964). "Chemotaxonomie der Pflanzen", Vol. III, Birkhäuser, Basel.

Hendrickson, J. B. (1959). *Tetrahedron*, **7**, 82.

Herout, V. (1966a). *Herba Hungarica*, **65**.

Herout, V. (1966b). *Planta Medica, Suppl.* **97**.

Herz, W. (1968). *In* "Recent Advances in Phytochemistry" (T. J. Mabry, ed.), Appleton Century Crofts, New York.

Herz, W., and Lakshmikantham, M. V. (1965). *Tetrahedron*, **21**, 1711.

Herz, W., and Santhanam, P. S. (1967). *J. org. Chem.* **32**, 507.

Herz, W., Mitra, R. B., Rabindran, K., and Viswanathan, N. (1962). *J. org. Chem.* **27**, 4041.

Herz, W., Chikamatsu, H., and Tether, L. R. (1966a). *J. org. Chem.* **31**, 1632.

Herz, W., Rajapa, S., Roy, S. K., Schmid, J. J., and Mirrington, R. N. (1966b). *Tetrahedron*, **22**, 1907.

Herz, W., Chikamatsu, H., Viswanathan, N., and Sudarsanam, V. (1967a). *J. org. Chem.* **32**, 682.

Herz, W., Rajapa, S., Lakshmikantham, M. V., Raulais, B., and Schmid, J. J. (1967b). *J. org. Chem.* **32**, 1042.

Herz, W., Sumi, Y., Sudarsanam, V., and Raulais, D. (1967c). *J. org. Chem.* **32**, 3658.

Hikino, H., Hikino, Y., and Yosioka, T. (1962). *Chem. pharm. Bull., Tokyo*, **10**, 641.

Hochmannová, J., and Herout, V. (1964). *Colln Czech. chem. Commun.* **29**, 2369.

Hochmannová, J., Novotný, L., and Herout, V. (1962). *Colln Czech. chem. Commun.* **27**, 2711.

Hoffmann, O. (1894). "Compositae" in Engler-Prantl, "Die natürlichen Pflanzenfamilien", Vol. IV/5, Engelmann, Leipzig.

Holzer, K., and Zinke, A. (1953). *Mfg. Chem.* **84**, 901.

Ishii, H., Tozyo, T., and Minato, H. (1965). *Tetrahedron*, **21**, 2605.

Kagan, H. B., Miller, H. E., Renold, W., Lakshmikantham, M. W., Tether, L. R., Herz, W., and Mabry, T. J. (1966). *J. org. Chem.* **31**, 1629.

Korte, F., and Bachmann, G. (1958). *Nature (London)*, **45**, 390.

Křepinsky, J., Motl, O., Dolejs, L., Novotný, L., Herout, V., and Bates, R. B. (1968). *Tetrahedron Lett.* 3315.

Kulkarni, G. H., Paul, A., Rao, A. S., Kelkar, G. R., and Bhattacharyya, S. C. (1961). *Tetrahedron*, **12**, 178.

Kulkarni, G. H., Bawdekar, A. S., Rao, A. S., Kelkar, G. R., and Bhattacharyya, S. C. (1963). *Perf. Record*, **54**, 303.

Kupchan, S. M., Aynehchi, L., Cassady, J. M., McPhail, A. T., Sim, C. A., Schnoes, H. K., and Burlingame, A. L. (1966). *J. Am. chem. Soc.* **88**, 3736.

Kushnir, L. E., and Kuzovkov, A. D. (1966). *Khim. prirod. Soedinenii*, 245.

Leonhardt, R. (1949). *Öst. bot. Z.* **96**, 293.

Mabry, T. J., Kagan, H. B., and Miller, H. E. (1966). *Tetrahedron*, **22**, 1943.

Merkel, D. (1966). *In* "Die Ätherischen Öle", (Gildemeister-Hoffmann) Vol. IIId, Akad. Verlag, Berlin.

Miller, H. E., Kagan, H. B., Renold, W., and Mabry, T. J. (1965). *Tetrahedron Lett.* 3397.

Miller, H. E., Mabry, T. J., Turner, B. L., and Payne, W. W. (1968). *Am. J. Bot.* **55**, 316.

Minato, H., and Horibe, I. (1965). *J. chem. Soc.* 7009.

Minato, H., Nosaka, S., and Horibe, I. (1964). *Proc. chem. Soc.* **120**.

Naves, Y. R. (1949). *Mfg. Chem.* **20**, 318.

Naya, K., Tagaki, I., Hayashi, M., Nakamura, S., and Kobayashi, M. (1967). Symp. Papers, "The Chemistry of Natural Products", p. 88, Kyoto.

Naya, K., and Tagaki, I. (1968). *Tetrahedron Lett.* 629.

Novotný, L., and Herout, V. (1962). *Colln Czech. chem. Commun.* **27**, 2462.

Novotný, L., and Herout, V. (1965). *Colln Czech. chem. Commun.* **30**, 3579.

Novotný, L., and Šorm, F. (1965). *In* "Beiträge zur Biochemie und Physiologie von Naturstoffen" (K. Mothes, ed.), p. 327, G. Fischer, Jena.

Novotný, L., Herout, V., and Šorm, F. (1962a). *Colln Czech. chem. Commun.* **27**, 1400.

Novotný, L., Jizba, J., Herout, V., and Šorm, F. (1962b). *Colln Czech. chem. Commun.* **27**, 1393.

Novotný, L., Jizba, J., Herout, V., Šorm, F., Zalkow, L. H., Hu, S., and Djerassi, C. (1963). *Tetrahedron*, **19**, 1101.

Novotný, L., Samek, Z., and Šorm, F. (1966a). *Colln Czech. chem. Commun.* **31**, 371.

Novotný, L., Toman, J., Starý, F., Marquez, A. D., Herout, V., and Šorm, F. (1966b). Phytochemistry, **5**, 1281.

Novotný, L., Herout, V., and Šorm, F. (1968a). *Colln Czech. chem. Commun.* (In Press.)

Novotný, L., Samek, Z., and Šorm, F. (1968b). *Colln Czech. chem. Commun.* (In Press).

Novotný, L., Toman, Z., and Herout, V. (1968c). *Phytochemistry* **7**, 1349.

Ognjanov, I., Ivanov, D., Herout, V., Horák, M., Plíva, J., and Šorm, F. (1958). *Colln Czech. chem. Commun.* **23**, 2033.

Panizzi, L. (1965). Private communication.

Patil, F., Lehn, J. M., Ourisson, G., Tamahashi, Y., and Takeoshi, T. (1965). *Bull. Soc. Chim. France*, 3085.

Rao, A. S., Kelkar, G. R., and Bhattacharyya, S. C. (1958). *Chem. Ind. (London)*, 1359.

Rao, A. S., Paul, A., Sadgopal, B. S., and Bhattacharyya, S. C. (1961). *Tetrahedron*, **13**, 319.

Ríos, T., de Vivar, A. R., and Romo, J. (1967). *Tetrahedron*, **23**, 4265.

Rivett, D. E. A., and Woolard, G. R. (1967). *Tetrahedron*, **23**, 2431.

Rodríguez-Hahn, L., Guzmán, A., and Romo, J. (1968). *Tetrahedron*, **24**, 477.

Romaňuk, M., Herout, V., and Šorm, F. (1958). *Colln Czech. chem. Commun.* **23**, 2188.

Romo, J., de Vivar, A. R., and Herz, W. (1963). *Tetrahedron*, **19**, 2317.

Romo, J., de Vivar, A. R., and Joseph-Nathan, P. (1967). *Tetrahedron*, **23**, 29.

Saleh, M. R. T. (1957). *Proc. pharm. Soc. Egypt, Sci. Ed.* **39**, 107 (cf. C.A. **52**, 20441, 1958).

Satoda, I., Yoshida, N., and Yoshii, E. (1959). *J. pharm. Soc. (Japan)*, **79**, 267.

Souček, M., Herout, V., and Šorm, F. (1961). *Colln Czech. chem. Commun.* **26**, 803.

Stahl, E. (1952a). *Pharmazie*, **7**, 863.

Stahl, E. (1952b). *Naturwiss.* **39**, 571.

Stoll, A., Morf, R., Rheiner, A., and Renz, J. L (1956). *Experientia*, **12**, 360.

Steelink, C., and Spitzer, J. C. (1966). *Phytochemistry*, **5**, 357.

Suchý, M. (1962). *Colln Czech. chem. Commun.* **27**, 2925.

Suchý, M., and Herout, V. (1961). *Colln Czech. chem. Commun.* **26**, 890.

Suchý, M., and Herout, V. (1962). *Colln Czech. chem. Commun.* **27**, 1510.

Suchý, M., Herout, V., and Šorm, F. (1960). *Colln Czech. chem. Commun.* **25**, 2777.

Suchý, M., Herout, V., and Šorm, F. (1962). *Colln Czech. chem. Commun.* **27**, 1905.

Suchý, M., Herout, V., and Šorm, F. (1965). *Colln Czech. chem. Commun.* **30**, 2863.

Suchý, M., Samek, Z., Herout, V., and Šorm, F. (1967a). *Colln Czech. chem. Commun.* **32**, 2016.

Suchý, M., Samek, Z., Herout, V., and Šorm, F. (1967b). *Colln Czech. chem. Commun.* **32**, 3934.

Suchý, M., Samek, Z., Herout, V., and Šorm, F. (1968a). *Colln Czech. chem. Commun.* **33**, 2238.

Suchý, M., Samek, Z., Snatzke, G., Herout, V., and Šorm, F. (1968b). *Colln Czech. chem. Commun.* (In Press.)

Sumi, M. (1958). *J. Am. chem. Soc.* **80**, 4869.

Šorm, F., and Dolejš, L. (1965). "Guaianolides and Germacranolides", Hermann, Paris.

Villiers, J. P., and Pachler, K. (1963). *J. chem. Soc.* 4989.

CHAPTER 8

Flavonoid Patterns in the Monocotyledons

E. C. BATE-SMITH

*Agricultural Research Council Institute of Animal Physiology,
Babraham, Cambridge, England*

I.	Introduction	167
II.	Flavonoid Patterns in the Dicotyledons	169
III.	Monocotyledons and Dicotyledons Compared	172
IV.	Flavonoids of Selected Monocotyledonous Families . . .	172
V.	Taxonomic and Phylogenetic Implications of the Flavonoid Patterns .	175
	References	177

I. INTRODUCTION

Much less is known of the chemistry of the monocotyledons than of the dicotyledons. The flavonoid constituents in this division of the angiosperms are therefore, best discussed in relation to what is known about the patterns which have been discerned in the much-better-known dicotyledons (Bate-Smith, 1962). A corresponding treatment of the monocotyledons, of which the present paper is a summary, has been published more recently (Bate-Smith, 1968).

In any discussion on a subject such as this, some orderly arrangement of the objects studied has to be decided upon. In the case of plants there are any one of a number of accepted systems of classification, and of these that of Engler seems to be the most widely known and used. This system however, is not by any means a fixed and definitive one; each edition from the original one in *Die natürlichen Pflanzenfamilien* (1887–) to the latest one in the 12th edition of the *Syllabus der Pflanzenfamilien* (1964) introduces some changes in arrangement. The most stable of these arrangements is that given in the 6th edition of Willis's *Dictionary of Flowering Plants and Ferns* (1951), and that is the one which I have followed in my discussion of the flavonoid patterns in both dicotyledons and monocotyledons (Bate-Smith, 1962, 1968).

Another way of arranging the families of plants is on a phylogenetic basis, and several authors have erected systems of classification on frankly phylogenetic principles. Three such systems relating to the monocotyledons are

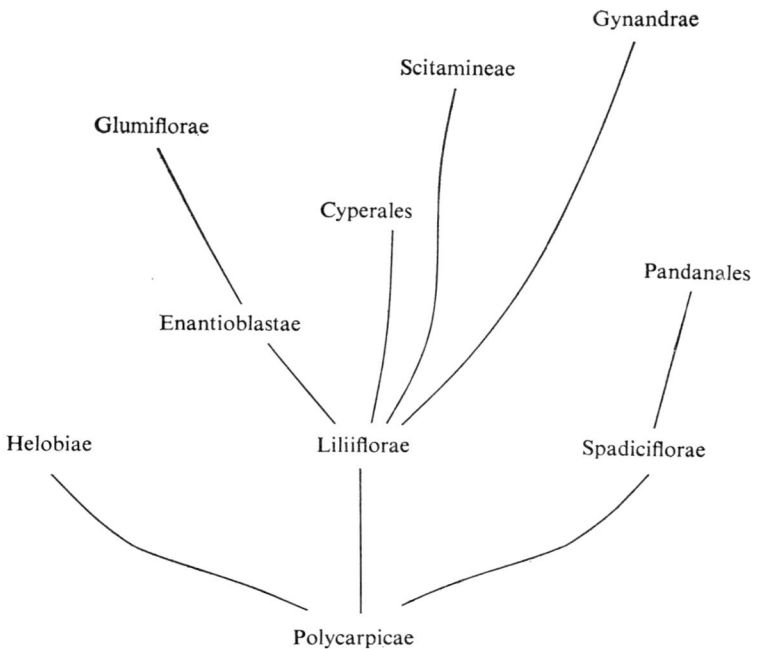

FIG. 1. Relationships in the Monocotyledoneae according to Wettstein. (From Hegnauer (1963). *Reproduced by permission of the author and of Birkhaüser-Verlag.*)

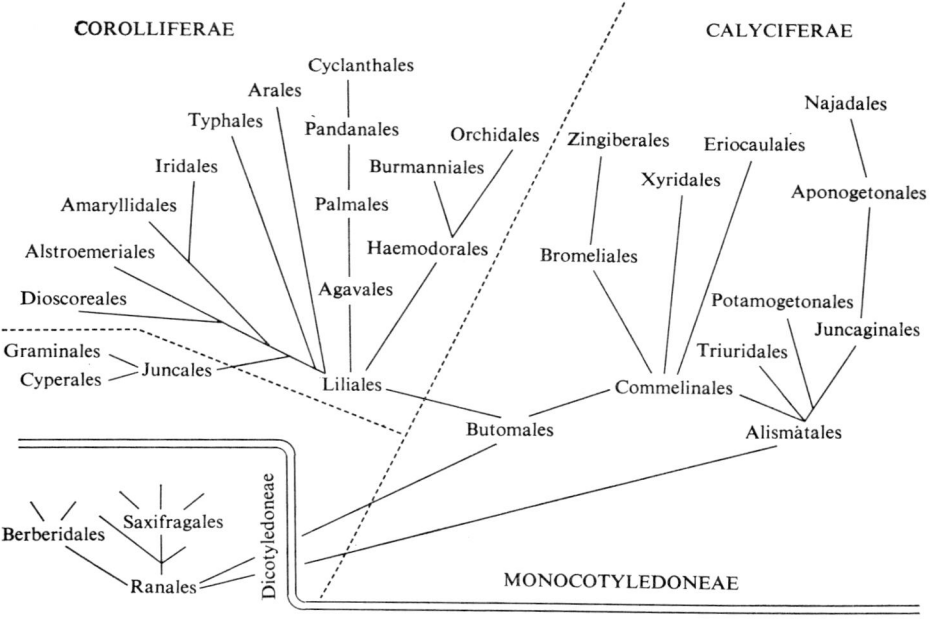

FIG. 2. Relationships in the Monocotyledonea according to Hutchinson. (From Hegnauer (1963). *Reproduced by permission of the author and of Birkhaüser-Verlag.*)

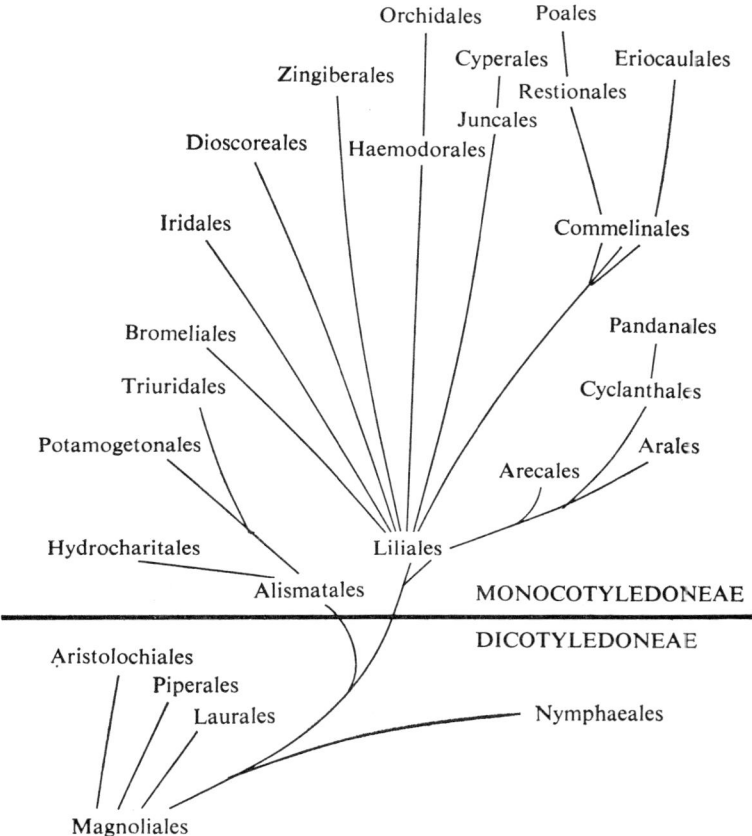

FIG. 3. Relationships in the Monocotyledoneae according to Takhtajan. (Frcm Hegnauer (1963). (*Reproduced by permission of the author and of Birkhaüser-Verlag.*)

shown in Figs. 1–3. Each of these presumes an origin of the monocotyledons by divergence from a dicotyledonous stock, usually Ranalian or pre-Ranalian. The subsequent evolutionary pathways, however, tend to differ. Since the flavonoid distribution in the dicotyledons seems to be able to contribute materially to the development of such concepts as these, I shall preface what I have to say about the distribution of these substances in the monocotyledons with a summary of what appears to be their phylogenetic significance in the dicotyledons.

II. FLAVONOID PATTERNS IN THE DICOTYLEDONS

In the dicotyledons, the three most commonly occurring flavonoid constituents are the flavonols, quercetin (I) and kaempferol (II), and the flavandiol,*

* It is probable that flavandiols always occur as dimers or higher polymers.

leucocyanidin (III). Frequently these are accompanied by the flavonol, myricetin (IV), and the flavandiol, leucodelphinidin (V), both of which have the vicinal trihydroxy grouping in the B ring. The flavonols are so widespread in their occurrence that their distribution is only of secondary taxonomic significance, but the presence or absence of leuco-anthocyanins* (denoted a and a_0, respectively) is highly significant taxonomically. Similarly the presence or absence (denoted b and b_0, respectively) of the *vic*-trihydroxy substituted

Quercetin (I)

Kaempferol (II)

Leucocyanidin (III)

Myricetin (IV).

Leucodelphinidin (V)

flavonoids is highly significant. There is much evidence to support the view that the a and b conditions are primitive to a_0 and b_0, but the a condition is also highly correlated with woodiness. Thus, nearly all the predominantly woody families from Casuarinaceae to Ebenaceae are type a, while nearly all the herbaceous families from Aristolochiaceae to Compositae are a_0. The b and b_0 conditions are not so directly correlated with woody habit, but are related to situations which are generally regarded as primitive and advanced,

* i.e., all colourless compounds which give anthocyanidins on treatment with hot mineral acid, regardless of structure.

respectively; but since the families of predominantly woody plants are also those generally regarded as primitive, there is a tendency for such families to be of the *b* type.

There are a number of other correlations which help to clarify the status of particular plants or families of plants. When flavonols are absent, their place may be taken by flavonoids in a more reduced condition, such as flavones, flavanones and chalcones. Thus, there are whole families in the Tubiflorae in which the flavonols are absent, or are only present in low concentrations, and are then accompanied by the above vicariant substances. One class of flavonoids which assumes special significance in the monocotyledons are the glycoflavones. These compounds appear to be more nearly related to the flavonols than to the flavones, and hence do not carry as much implication of evolutionary advancement as the absence of flavonols would otherwise imply. Further light is shed on the possible phylogenetic situation in the dicotyledons by the pattern of distribution of the commonly occurring substituted cinnamic acids. Those regularly present in woody plants are caffeic (VI) and *p*-coumaric (VII) acids, whereas in herbaceous plants, ferulic (VIII) and sinapic (IX) acids

Caffeic Acid (VI)

p-Coumaric Acid (VII)

Ferulic Acid (VIII)

Sinapic Acid (IX)

more commonly accompany *p*-coumaric acid. Ferulic acid, it will be noted, is the 3-*O*-methylated derivative of caffeic acid, while sinapic acid is the 3,5-di-*O*-methylated derivative of 3,4,5-tri-hydroxycinnamic acid (X), not itself so far reported to occur in plants. Where by analogy the latter compound might be expected to occur, another acid, ellagic acid (XI) is often found.

3,4,5-Trihydroxycinnamic Acid (X)

Ellagic Acid (XI)

The occurrences of these phenolic acids, associated in the cases of caffeic and ellagic acid with the presence of flavonols and flavandiols, and in the cases of ferulic and sinapic acids with their absence, reinforce the indications provided by the flavonoid constituents (I–V) themselves regarding the evolutionary status of particular plants or families, a question discussed earlier by Harborne (Harborne, 1966).

III. Monocotyledons and Dicotyledons Compared

The flavonoid patterns found in monocotyledons and dicotyledons do not differ in any essential respect, nor, with one conspicuous exception, do those of the hydroxy- and methoxy acids just mentioned. The exception is ellagic acid, which has never so far been found in the monocotyledons. It has also to be borne in mind that the monocotyledons are overwhelmingly herbaceous: there is only one woody family, the Palmae, and elsewhere a few woody plants occur in the Liliaceae and, if bamboos are so regarded, in the Gramineae. The effect of woody habit on flavonoid distribution, which is often so decisive in the dicotyledons, is therefore less apparent in the monocotyledons. Moreover, the taxonomic guidelines are themselves still in dispute. While the absence of ellagic acid deprives us of a chemotaxonomic guideline which shows promise of being exceptionally helpful in the dicotyledons, this can in itself prove useful when, later on, it comes to discussing the phylogenetic relationships of these two major plant divisions.

Although there is a poverty of woody species in the monocotyledons, there is a correspondingly large representation of aquatic or moisture-loving species. Many of these contain leuco-anthocyanins,* some of the flavandiol type found in the dicotyledons, but others which differ in their properties and may therefore not be flavandiols. Since however, there is no certain way at present of deciding whether this is the case, leuco-anthocyanins will for the time being, be regarded as having the same taxonomic and phylogenetic significance in the monocotyledons as they appear to have in the dicotyledons; i.e. they are an indication of evolutionary primitiveness in the plants that possess them as compared with those that do not.

IV. Flavonoids of Selected Monocotyledonous Families

Of the forty-five families of monocotyledons recognized by Engler (cf. Willis, 1951), I have examined members of thirty-seven. The first ten of these, from Typhaceae to Hydrocharitaceae (see Table I), are mostly families of aquatic plants, and six of them contain leucocyanidin,† the exceptions being Pandanaceae, Potamogetonaceae, Scheuchzeriaceae and Butomaceae. Only four families contain flavonols; luteolin (XII) is frequently present, sometimes,

---

* See footnote on p. 170.
† Not necessarily (III).

TABLE I

Summary of the families of the Monocotyledoneae

Typhaceae—Hydrocharitaceae (10 families)
(Triuridaceae)
Gramineae, Cyperaceae
Palmae, (Cyclanthaceae)
Araceae—Stemonaceae (17 families)
Liliaceae—Dioscoreaceae—Iridaceae (7 families)
Musaceae—Marantaceae (4 families)
(Burmanniaceae), Orchidaceae

as in *Zostera* in the Potamogetonaceae and *Stratiotes* in the Hydrocharitaceae, in very large amounts. These are all small families, so they do not lend themselves to individual analysis, but taken as a whole they exemplify very well the overall distribution of flavonoids in the monocotyledons: association of leuco-anthocyanin with aquatic or moisture-loving habit, lacking in flavonols but rich in flavones; and, in these particular families, complete absence of trihydroxyphenyl constituents (i.e. b_0).

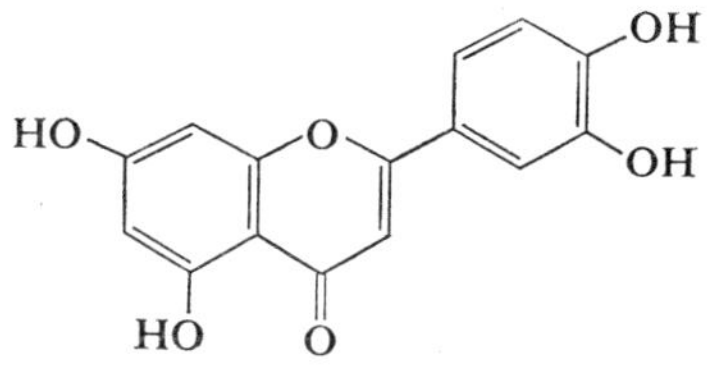

Luteolin (XII)

The next family, the Gramineae, except for the extreme rarity of leuco-anthocyanins, has a similar pattern. Leucocyanidin has, in fact, been reported only twice in the whole family: once in the endosperm of barley, and once in the seed-coat of a single variety of *Sorghum vulgare* (Bate-Smith and Rašper, 1968). The Andropogoneae, to which tribe *Sorghum* belongs, have, however, large amounts of a quite novel leuco-anthocyanin based on a flavan-4-ol structure and yielding luteolinidin (XIII) when heated with mineral acid. Flavonols are also very rare, but glycoflavones are commonly present, probably occurring more often than luteolin (XII). Caffeic acid (VI) is seldom present,

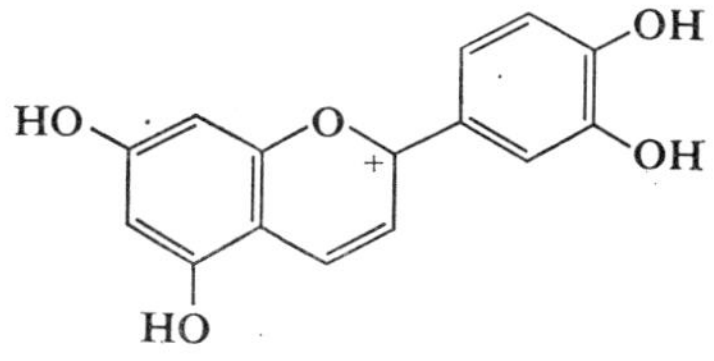

Luteolinidin (XIII)

but ferulic acid (VIII) is invariably found. This is, therefore, a very uniform family with few primitive and many advanced chemotaxonomic features.

The Cyperaceae, usually regarded as closely related to the Gramineae, are divided into three sub-families. The Caricoideae are very similar in flavonoid content to the Gramineae, but the Scirpoideae differ completely, being rich in flavonols and leuco-anthocyanins, one of the species examined having, in fact, the trihydroxy substituted leucodelphinidin (V). The third sub-family has not been examined as yet.

All species of the wholly woody Palmae examined contain leucocyanidin (III) together with other leuco-anthocyanins giving anthocyanidins of higher R_f value than cyanidin (i.e. not delphinidin). Flavonols are only feebly represented, but caffeic acid (VI) is usually present, together with the methoxy-cinnamic acids (VIII and IX).

Of the seventeen families lying between the Palmae and the Liliaceae in the Englerian system, thirteen have been examined, and these seem to fall into two chemotaxonomic groups, one resembling those which are predominantly aquatic and the other the grasses. The former comprise the Araceae, Lemnaceae, Pontederiaceae, Philydraceae and perhaps the Bromeliaceae and Commelinaceae; the latter the Flagellariaceae, Restionaceae, Mayacaceae, Xyridaceae, Eriocaulaceae, Cyanastraceae and Juncaceae. These lines again seem to follow to some extent the moist- and dry-habitat tendency, but the numerous families represented are mostly small in numbers of species and are too heterogeneous for any clear indications to be sought. It is interesting, however, that, here and there, leucodelphinidin has been found and in *Eriocaulon* species a flavonol that is either quercetagetin (XIV) or gossypetin (XV) is usually present, the only instance of the occurrence of such a compound in the monocotyledons.*

Quercetagetin (XIV)   Gossypetin (XV)

Because of their size and diversity, the Liliaceae and Iridaceae provide the best opportunity for considering the taxonomic implications of the flavonoid patterns in the monocotyledons. In the Smilacoideae, the Liliaceae possess one woody tribe, and these do, in fact, have the regular woody pattern of flavonoids. Several of the other tribes, e.g. the Allioideae and the Lilioideae have plentiful flavonols and very occasionally some members have leucocyanidin and leucodelphinidin (these, incidentally, are usually Southern hemisphere plants). Ferulic and sinapic acids are much more frequently present than is caffeic acid.

The Iridaceae, entirely herbaceous, nevertheless have some members,

* See note added in proof p. 177.

especially in the genus *Iris*, with leuco-anthocyanins, usually leucodelphinidin as well as leucocyanidin. Some *Iris* and *Crocus* species also contain flavonols, and again myricetin (IV) is present along with quercetin (I) and kaempferol (II) —a most unusual occurrence in the monocotyledons. Most of the members of the family, however, contain neither flavonols nor flavandiols, thus closely resembling the Liliaceae in their flavonoid patterns.

The Dioscoreaceae, Musaceae, Zingiberaceae, Cannaceae and Marantaceae, all herbaceous, have similar patterns to that in the genus *Iris* described above. In these families ferulic acid is almost invariably present, but so also is caffeic acid, so that the "woody" flavonoid pattern is here completely reconciled with a "herbaceous" pattern of subsidiary constituents.

The evolutionary position assigned to these families in the Englerian and other systems is almost always an advanced one, yet they would be accorded the most primitive status if the principles which apply to the distribution of flavonoids in the dicotyledons are applicable. The Dioscoreaceae, for instance, are accorded the most advanced status in the Liliiflorae, but, along with some members of the Iridaceae, they have the most primitive chemotaxonomic characters. It is therefore interesting to note that Lawton and Lawton (1967) have recently reported that some *Dioscorea* species possess dicotyledonous embryos, suggesting that the position of this family as a member of the monocotyledons might even have to be reconsidered!

There is only one family left to deal with, that is the Orchidaceae. Here, as in the Gramineae, there are a few species which contain leucocyanidin (all in the tribe Dendrobiae), but flavonols are almost completely absent. Only two species out of the twenty-three so far examined contain large amounts of caffeic acid. For the most part the phenolic constituents are unusual and unidentified. This, therefore, can confidently be designated an advanced family, a view with which I think few botanists would disagree.

V. TAXONOMIC AND PHYLOGENETIC IMPLICATIONS OF THE FLAVONOID PATTERNS

Without going into greater detail, it is impossible to say much more about the taxonomic implications of the findings outlined above. Taxonomy tends to be concerned with minutiae, but phylogeny allows a much broader treatment. I have studied two genera of the Iridaceae in much greater detail, and have found that in *Iris*, where the morphological characters enable clear-cut divisions to be made, the chemical characters closely follow these divisions, but in *Crocus*, where no such clear-cut divisions have been identified, the chemical characters are equally unclear. There are indications, however, of a correlation of chemical characters with *geographical distribution* of species in *Crocus*, and the same can be said of *both* morphological and chemical characters in *Iris*. These considerations are discussed in detail in the paper already referred to (Bate-Smith, 1968).

If we look a little further into the case of *Crocus* as an example of taxonomic treatment of a genus, we find that the dichotomy proposed by Maw, the monographer of the genus (1886), into sections Involucrati and Nudiflorae is followed by a sub-division of each section into subsections Fibromembranacei and Reticulati (referring to the tunic of the bulb). In other words, the "dichotomy" is not a dichotomy at all. This kind of situation seems often to obtain in the monocotyledons; relationships proposed on one set of characters can be contraindicated by another set, and instead of a dichotamous (or "ramose") taxonomic structure one finds a "reticulate" structure, with horizontal links between vertical components. This may, in fact, be the true state of affairs, but if so it is one which allows no useful progress to be made in identifying the lines along which evolution has progressed.

The origin of the monocotyledons is "one of the great unsolved problems of phylogeny" (Willis, 1951). In the absence of an agreed taxonomy of the class, it is difficult to see how the origin can be traced back to particular ancestral progenitors, but something can be deduced from the chemotaxonomic data as to the likeliest point of departure from the angiospermous line, and the likeliest candidates for qualifying as ancestral progenitors. A clue is provided by the absence of ellagitannins from the monocotyledons. These substances are complex derivatives of hexahydroxydiphenic acid (XVI) (the ring-opened form of the dilactone ellagic acid (XI)) and are present in a large number of

Hexahydroxydiphenic Acid (XVI)

dicotyledonous orders, among which the Rosales appear to occupy a central position, but are absent from an equally large number of orders, exemplified especially by the Ranales. It appears an inescapable conclusion that the ellagitannins appeared at a particular point in the evolution of the dicotyledons, and it is intriguing to speculate that this was a *unique* event, and that all the plants which possess these tannins derive from the unique, "hapactic" occurrence (the term is a new one, from ἄπαξ, once only). If this suggestion is accepted, it follows that the monocotyledons derived from the angiospermous stock before this event, and are more closely related to Ranalian than to Rosalian ancestors.

As regards the candidates for ancestral honours, reference can usefully be made to Lowe's study of the phylogeny of monocotyledons (Lowe, 1961). On the basis of a kind of numerical taxonomy introduced by Sporne (1949), she identifies families of the Liliiflorae as the most primitive and the aquatic families (Table I 1–10) as among the most advanced. The Dioscoreaceae, which I have indicated as rather strong candidates for primitive status, are among the lowest, and one of the genera, *Petermannia* (given a family to itself by Hutchinson) is rated lowest of all! Lowe supports Sargant's (1903, 1904) view that the link with the primitive angiospermous stock is through the Liliaceae, and that the Alismataceae (the general favourites for this role) are secondarily aquatic. It is interesting to see in Figs. 1–3 how three different systematists have interpreted this question.

Lowe's analysis also allows us to come to some conclusions about the significance of leuco-anthocyanins in the aquatic monocotyledons. In these non-woody plants, these compounds clearly do not have the significance assigned to them in the dicotyledons, and they may not, in fact, be chemically or biosynthetically homologous with the leuco-anthocyanins of woody taxa. One distinction is that the pigment produced when such plants are heated with mineral acid is a clear rose colour, unaccompanied by any trace of a coagulum of "phlobaphene", and this may give a clue as to how to establish a distinction between the two types of compound—a distinction which is obviously important both taxonomically and phylogenetically.

Finally, it has to be remembered that *all* present-day plants have their roots in the distant past. Many of the families discussed above have one or more genera with a "primitive" flavonoid pattern, however "advanced" that of the majority of its members may be. The ancestry of each of these families is therefore traceable back *through these genera* to the primitive angiospermous stock, and to be able to do this is something for which a chemotaxonomic approach may perhaps claim some credit.

REFERENCES

Bate-Smith, E. C. (1962). *J. Linn. Soc. (Bot.)* **58**, 95.
Bate-Smith, E. C. (1968). *J. Linn. Soc. (Bot.)* **60**, 325.
Bate-Smith, E. C., and Rašper, V. (1968). *J. Fd Sci.* (In Press.)
Engler, A. (1964). "Syllabus der Pflanzenfamilien", 12th edition (H. Melchior, ed.), II Band, Angiospermae. Bornträger, Berlin.
Harborne, J. B. (1966). *In* "Comparative Phytochemistry" (T. Swain ed.), Academic Press, London and New York.
Hegnauer, R. (1963). "Chemotaxonomie der Pflanzen", Band 2. Monocotyledoneae. Birkhaüser Verlag. Basel and Stuttgart.
Lawton, J. R. S., and Lawton, J. R. (1967). *Proc. Linn. Soc. Lond.* **178**, 153.
Lowe, J. (1961). *New Phytol.* **60**, 355.
Maw, G. (1886). "The genus *Crocus*", Dulau.
Sargant, E. (1903). *Ann. Bot.* **17**, 1.
Sargant, E. (1904). *Bot. Gaz.* **37**, 325.
Sporne, K. R. (1949). *New Phytol.* **48**, 259.
Willis, J. C. (1951). "A dictionary of flowering plants and ferns", 6th edition, Cambridge University Press.

NOTE ADDED IN PROOF

The flavonol in *Eriocaulon* has since been identified as quercetagetin XIV (Bate-Smith and Harborne, unpublished) but the isomeric gossypetin XV has also been found in the monocotyledons for the first time, in *Restio Restionaceae* (Harborne and Clifford, unpublished).

CHAPTER 9

Metabolism of Cinnamic Acid and its Derivatives in Basidiomycetes

G. H. N. TOWERS

Department of Botany, University of British Columbia, Vancouver, Canada

I. Introduction	.	179
II. Phenylalanine and Tyrosine Ammonia Lyases in Basidiomycetes .	.	181
III. Metabolism of Phenylalanine in Basidiomycetes	.	182
IV. Degradation of Phenylalanine and Tyrosine in Basidiomycetes	.	183
V. Formation of more complex Cinnamoyl Derivatives including Lignins	.	187
VI. Comparison of Cinnamate Metabolism in Basidiomycetes and in Higher Plants	.	189
References	.	190

I. INTRODUCTION

Fungi comprise a large assemblage of organisms which share, with vascular plants and bacteria, an ability to synthesize a variety of secondary metabolites. Most of the phenols which they produce are not the typical natural products of higher plants but resemble instead the secondary products of bacterial metabolism. Basidiomycetes, however, appear to be exceptional.

A survey of the literature shows that cinnamic acid and its derivatives have been identified in Basidiomycetes. In many cases, the sporophores or fruiting bodies have been the source of material and since these structures are often contaminated with plant debris there is reason to doubt a fungal origin for the isolated compound. Moreover, it could be assumed that the small amounts of compounds usually located in fungal fruiting bodies arrived there by translocation from the host plant, especially in those wood-destroying Basidiomycetes which degrade lignins to simpler phenolics. Similarly, the large number of compounds related to cinnamic acid which have been reported to occur in the uredospores of wheat rust (*Puccinia graminis* var. *tritici*) (Van Sumere *et al.*, 1957) may represent absorbed materials of the host rather than products of fungal metabolism. Compounds related to cinnamic acid which have been discovered in Basidiomycetes are listed in Table I.

Nord and Vitucci (1947) reported that the Basidiomycete, *Lentinus lepideus*,

TABLE I

Occurrence of cinnamic acid and its derivatives in Basidiomycetes

Compound	Organism	Reference
Cinnamaldehyde	*Stereum subpileatum*	Birkinshaw *et al.*, 1957
Cinnamic acid	*Ceratocystis fimbriata*	Birkinshaw *et al.*, 1957;
	Stereum subpileatum	Kubota and Naya, 1954
Methyl cinnamate	*Lentinus lepideus*	Birkinshaw and Findlay, 1940.
Methyl *p*-coumarate	} *Lentinus lepideus*	Nord and Vitucci, 1947
Methyl *p*-methoxycinnamate		
Caffeic acid	*Boletus scaber*	Edwards and Elsworthy, 1967
p-Coumaric acid ⎤		
o-Coumaric acid ⎟		
Ferulic acid ⎟	*Puccinia graminis* var.	van Sumere *et al.*, 1957
⎬	*tritici* (spores)	
Caffeic acid ⎟		
Coumarin ⎟		
Umbelliferone ⎦		
Chlorogenic acid	*Fistulina hepatica*	Paris *et al.*, 1960
	Amanita citrina	
Phenylcrotonaldehyde	*Phallus impudicus*	List and Freund, 1966
Hispidin (XXIV)	*Polyporus hispidus*	Bu'Lock *et al.*, 1962;
	P. schweinitzii	Bu'Lock and Smith, 1961; Ueno *et al.*, 1964
Cortisalin (XXV)	*Corticium salicinum*	Gripenberg, 1952; Marshall and Whiting, 1957

accumulated methyl *p*-methoxycinnamate (V, methyl ester; (Fig. 2)) when grown on glucose, xylose and ethanol as sole carbon source. Labelled methyl *p*-methoxycinnamate was obtained when the organism was grown in media containing glucose-^{14}C (Shimazono *et al.*, 1958). This marked the first time that the *de novo* synthesis of a cinnamyl compound in a fungus was demonstrated.

These compounds usually have been considered to be characteristic secondary metabolites of higher plants, as well as of the Bryophyta, Lycopodophyta, Psilophyta and Equisetophyta. They have not been found in bacteria, algae or animals. The compounds shown in Table I are but a few examples of the wide range of plant phenolic substances based on or derived from a phenylpropane carbon skeleton. This phenylpropane unit is used in many synthetic processes in higher plants.

The precursors of phenylpropane units have been shown, in most instances, to be the amino acids phenylalanine (I) and tyrosine (particularly the former) which are deaminated to the corresponding cinnamic acids. The ammonia lyases required for catalysing the deamination are probably ubiquitous in green land plants (see Young *et al.*, 1966 and references cited therein). Cinnamic (II) and *p*-coumaric acid (III) in turn may be esterified, hydroxylated, glycosylated or *O*-methylated. They may serve as starters in polyketide biosynthesis leading to the production of flavonoids. The corresponding alcohols, *p*-coumaryl, coniferyl and sinapyl alcohols are almost certainly the precursors of lignins.

The production of derivatives of cinnamic acid is important in the carbon economy of land plants and, from a comparative point of view, it is of interest to determine whether this is also a feature of Basidiomycete metabolism.

II. PHENYLALANINE AND TYROSINE AMMONIA LYASES IN BASIDIOMYCETES

An initial survey by Power *et al.* (1965) revealed that twelve species of wood-destroying Basidiomycetes contained ammonia lyases which deaminated phenylalanine and tyrosine. A further survey (Bandoni *et al.* 1968) showed that a number of other types of Basidiomycetes as well as some members of the Fungi Imperfecti yielded these enzymes. Even the uredospores of *Puccinia graminis* (A. O. Jackson, personal communication) contain tyrosine ammonia lyase (TAL).

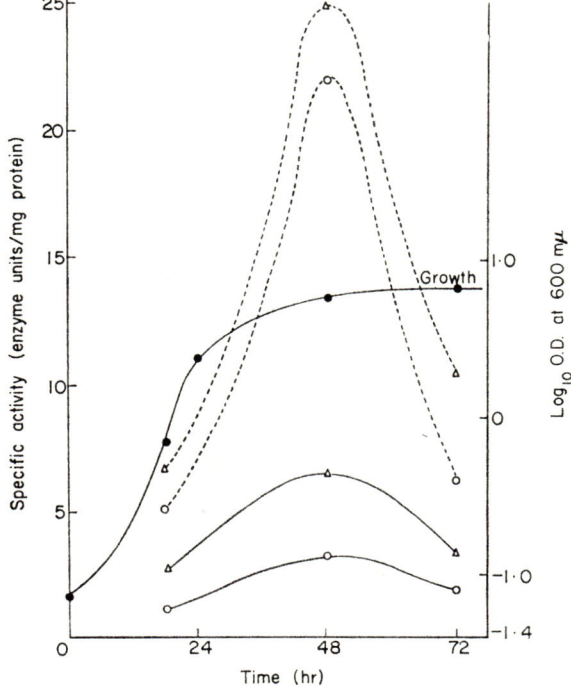

FIG. 1. Changes in phenylalanine and tyrosine ammonia lyase concentrations in relation to substrate in *Sporobolomyces roseus*. △———△ PAL activity, glucose alone; ○———○ TAL activity, glucose alone; △– – –△ PAL activity, glucose + 0·05% tyrosine; ○ – – – ○ TAL activity, glucose + 0·05 tyrosine. ●———● growth curve.

The enzymes from species of *Rhodotorula* (Ogata *et al.*, 1966, 1967a,b), from *Ustilago hordei* (Subba Rao *et al.*, 1967) and from *Sporobolomyces roseus* (Camm and Towers, unpublished results) have been studied in some detail. Enzyme activity increases during the later stages of the logarithmic phase of growth in *Sporobolomyces*, *Rhodotorula* and *Ustilago*. There is some evidence that both constitutive and inducible forms of phenylalanine ammonia lyase (PAL) occur in *Rhodotorula* species (Ogata *et al.*, 1967a). In *Rhodotorula glutinis*, *R. texensis* and in *Sporobolomyces* (Camm and Towers, unpublished) all imperfect fungi, it has not been possible so far to obtain a PAL preparation which does not also catalyse the deamination of tyrosine. The ratio of the two activities (PAL/TAL) does not vary sufficiently during the growth of *Sporobolomyces* cultures to indicate whether one or two enzymes exist (Fig. 1). *Ustilago hordei*, by contrast, does not exhibit any TAL activity (Subba Rao *et al.*, 1967). Purified preparations of PAL from this organism are completely inactive with tyrosine and, although the enzyme is inhibited by cinnamic acid, it differs from the enzymes of barley (Koukol and Conn, 1961) or sweet potato (Minamikawa and Uritani, 1965) in that it is not inhibited by tyrosine or *p*-coumaric acid.

A number of yeast-like Basidiomycetes and imperfect fungi such as *Rhodotorula* are readily cultured and offer excellent material for detailed studies of such enzymes.

III. METABOLISM OF PHENYLALANINE IN BASIDIOMYCETES

The occurrence of enzymes catalysing the production of cinnamic (II) and *p*-coumaric (III) acids in a group of organisms which can obtain these compounds from higher plants quite readily by parasitism or saprophytism is remarkable. It suggests that this is an important or basic metabolic feature of certain fungi. Tracer studies with growing cultures of the Basidiomycete, *Lentinus lepideus*, which produces a "brown" rot in wood, showed that it incorporated carbon from D-glucose and L-phenylalanine into isoferulic acid (VI) whereas L-tyrosine and acetate were poor precursors (Power *et al.*, 1965). It is clear therefore that the enzyme, PAL, which occurs in the organism is functional. TAL activity can also be detected in *Lentinus* but apparently the carbon of tyrosine is not diverted into these particular metabolites.

Earlier studies in Nord's laboratory on the biosynthesis and inter-relationships of a group of closely related derivatives of cinnamic acid, namely, the methyl esters of *p*-coumaric (III), *p*-methoxycinnamic (V) and isoferulic acids (VI) in *L. lepideus* have been reviewed by Schubert (1965). Methyl *p*-coumarate-carboxyl-^{14}C was converted to methyl *p*-methoxycinnamate without scrambling of the label (Shimazono, 1959). Power *et al.* (1965) showed that the main products of *p*-coumarate-^{14}C metabolism were caffeic (VII) and phloretic acids (IV). The known transformations of cinnamyl compounds in this organism are summarized in Fig. 2.

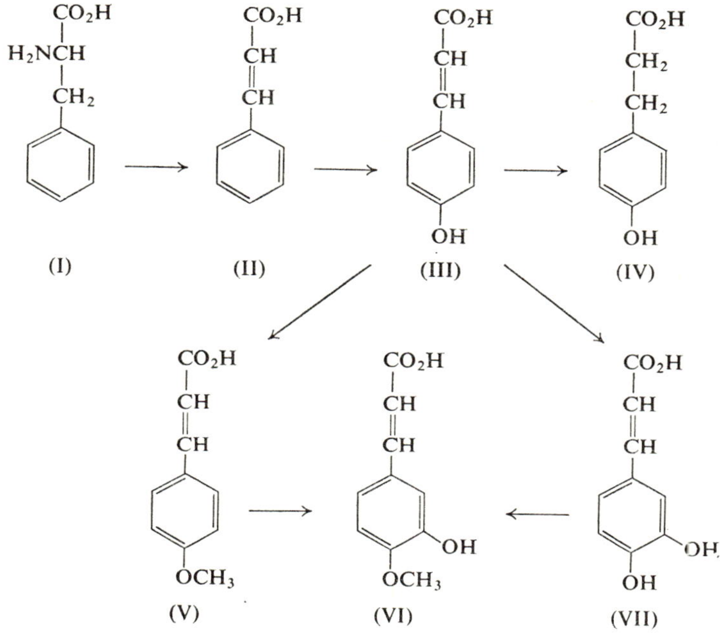

FIG. 2. Phenylpropanoid compounds formed from phenylalanine in *Lentinus lepideus*. A number of these accumulate in the medium as methyl esters. (Based on work of Shimazono (1959) and of Power *et al*. (1965).) (I) Phenylalanine; (II) cinnamic acid; (III) *p*-coumaric acid; (IV) phloretic acid; (V) *p*-methoxycinnamic acid; (VI) isoferulic acid; (VII) caffeic acid.

L-Tyrosine-^{14}C was not converted to cinnamyl derivatives in *Lentinus* but rather to *p*-hydroxyphenylacetic acid (XII) (Power *et al*., 1965). In *Sporobolomyces roseus*, on the other hand, tyrosine-^{14}C was converted to *p*-coumarate and, in addition, *m*-coumaric acid was identified as a product of *m*-hydroxyphenylalanine metabolism (Moore *et al*., 1968).

IV. DEGRADATION OF PHENYLALANINE AND TYROSINE IN BASIDIOMYCETES

In the micro-organisms which have been studied phenylalanine and tyrosine are degraded via homogentisic acid (IX) with the initial formation of the corresponding *p*-hydroxyphenylpyruvic acid (VIII), a route characteristic of mammals (Evans, 1963; Jones *et al*., 1952; Kluyver and van Zijn, 1951; Meister, 1965; Rogoff, 1961; Trecanni, 1963). This pathway is shown in Fig. 3. Evidence for the operation of such a pathway from tyrosine has not been found so far in Basidiomycetes.

There is suggestive evidence, however, for a modified pathway leading from tyrosine to gentistic acid in these fungi. Crowden's (1967) studies with *Polyporus tumulosus* indicates that tyrosine is converted through *p*-hydroxyphenyl-

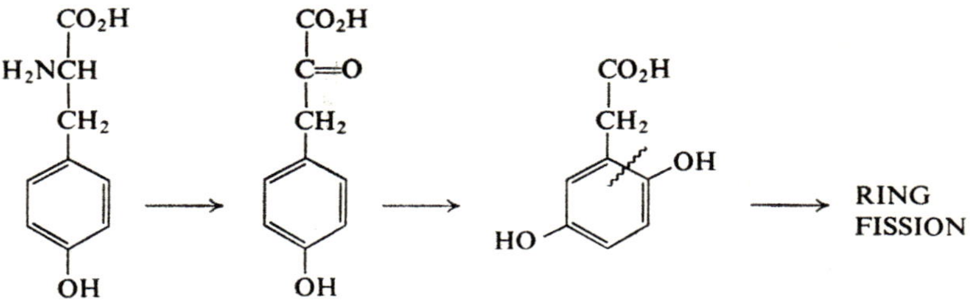

FIG. 3. Initial steps in the degradation of tyrosine in bacteria and mammals. (VIII) *p*-Hydroxyphenylpyruvic acid; (IX) homogentisic acid.

pyruvic (VIII), homogentisic (IX), 2,5-dihydroxymandelic (X), 2,5-dihydroxybenzoylformic (XI) and gentisic acids (XII) to CO_2 (Fig. 4a). The formation of gentisic acid-^{14}C from shikimic acid-U-^{14}C rather than from acetate-^{14}C was also demonstrated with this organism. Due to the long duration of the experiment (21 days) the results cannot be considered conclusive. It is of interest that in the Ascomycetes, *Penicillium patulum* and *P. urticae*, gentisic acid (XII) appears to be a product of polyketide metabolism and not of the shikimic acid pathway (Bassett and Tanenbaum, 1958; Bu'Lock and Ryan, 1958; Gatenbeck and Lonroth, 1962). Mycelial homogenates of *P. tumulosus* were found to shorten oxidatively the side chain of 2,5-dihydroxyphenylacetic (homogentisic) acid (IX) and 2,5-dihydroxybenzoylformic acid (XI). The oxidation of homogentisic acid by an enzyme preparation from *Poria subacida* has been reported although the product, presumed to be 2,5-dihydroxybenzoylformic acid (XI), was not identified (Fukuzumi, 1962). The oxidative conversion of homogentisic to gentisic acid has been shown enzymically with rabbit liver preparations (Ichihara *et al.*, 1956; Sakamoto *et al.*, 1958).

A second route for the degradation of phenylalanine and tyrosine in Basidiomycetes may be through the corresponding keto acids to phenylacetic and *p*-hydroxyphenylacetic acids (XIII) (see Fig. 4b). In *Schizophyllum commune* phenyllactic, phenylacetic and *o*-hydroxyphenylacetic acids were identified as products of phenylalanine-^{14}C metabolism (Moore and Towers, 1967). In *Polyporus tumulosis* it would appear that tyrosine is converted to the 4-hydroxy and 3,4-dihydroxy-derivatives of phenylpyruvic, phenylacetic, mandelic, benzoylformic and benzoic acids (Crowden, 1967). The relative importance, if any, of this route or the previous one in fungi has not been assessed carefully and they should be considered as no more than suggestive.

It is of considerable interest that evidence for a route leading to the degradation of phenylalanine through cinnamic acid has been found recently in *Schizophyllum commune* (Moore and Towers, 1967), *Sporobolomyces roseus* (Moore *et al.*, 1968) and *Ustilago hordei* (Moore *et al.*, 1967). Growing cultures of *Schizophyllum* not only convert phenylalanine to cinnamic acid but also

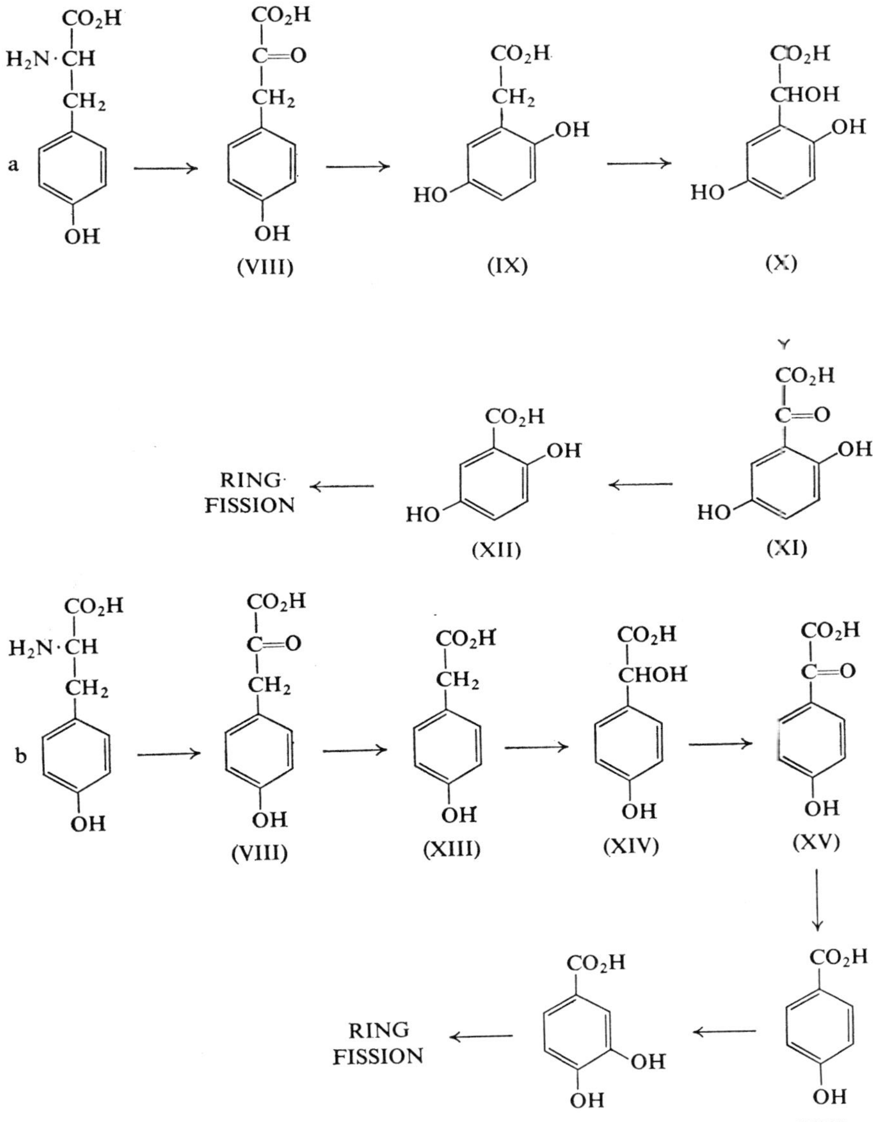

FIG. 4a. A pathway for the degradation of tyrosine in *Polyporus tumulosus* and its alternative 4b (after Crowden, 1967). (VIII) *p*-hydroxyphenylpyruvic acid; (IX) homogentisic acid; (X) 2,5-dihydroxymandelic acid; (XI) 2,5-dihydroxybenzoylformic acid; (XII) gentisic acid; (XIII) *p*-hydroxyphenylacetic acid; (XIV) *p*-hydroxymandelic acid; (XV) *p*-hydroxybenzoylformic acid; (XVI) *p*-hydroxybenzoic acid; (XVII) protocatechuic acid.

produce $^{14}CO_2$ when administered ring-labelled phenylalanine, cinnamic and benzoic acids. Although $L(-)$-β-phenyllactic-^{14}C-acid and phenylacetic-^{14}C acid are also products of phenylalanine the main route to CO_2 appears to follow the sequence cinnamate, benzoate, p-hydroxybenzoate and protocatechuate. Similar results have been obtained with *Ustilago* and *Sporobolomyces* except that, in these organisms phenyllactate and phenylacetate are not produced. Protocatechuic acid (XVII) is the terminal aromatic compound formed as a result of metabolism of cinnamic, p-coumaric, caffeic, benzoic and p-hydroxybenzoic acids. The degradation of phenylalanine in *Schizophyllum*

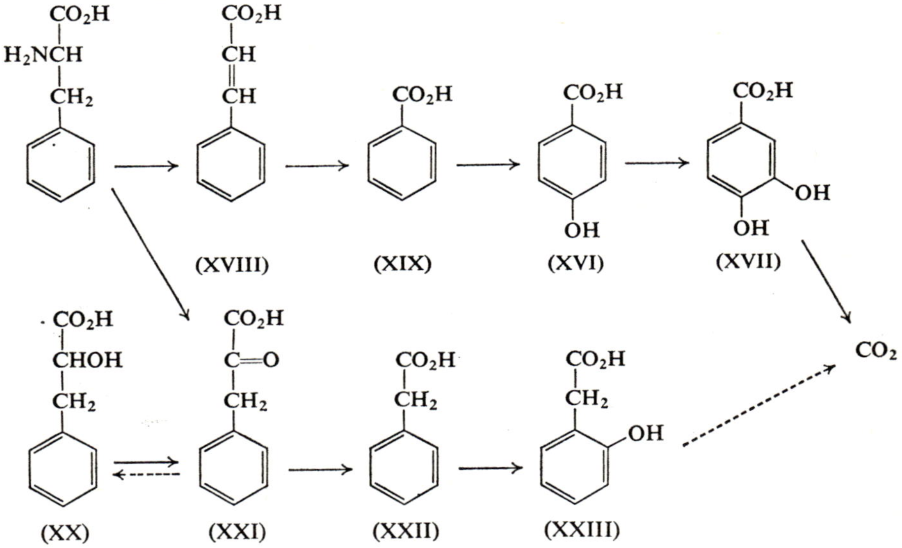

FIG. 5. Pathways for the degradation of L-phenylalanine in *Schizophyllum commune*. (XVIII) Cinnamic acid; (XIX) benzoic acid; (XVI) p-hydroxybenzoic acid; (XVII) protocatechuic acid; (XX) phenyllactic acid; (XXI) phenylpyruvic acid; (XXII) phenylacetic acid; (XXIII) o-hydroxyphenylacetic acid.

commune is shown in Fig. 5 and time-course graphs showing the disappearance of substrate and the formation of protocatechuic acid in *Sporobolomyces roseus* are shown in Figs. 6.

These organisms do not appear to hydroxylate cinnamic or p-coumaric acids. *Lentinus lepideus*, on the other hand, is capable of producing p-coumarate from cinnamate, and caffeate from p-coumarate (Power *et al.*, 1965). Benzoic acid, in contrast to cinamic acid, is readily hydroxylated in the *para* position by all these organisms. m-Hydroxybenzoic acid, again, is not hydroxylated to give protocatechuic acid but accumulates in *Sporobolomyces* when m-coumaric acid is the substrate (Moore *et al.*, 1968). o-Coumaric acid is metabolized to 4-hydroxycoumarin.

Although tyrosine appears to follow a similar route as does phenylalanine, i.e., through *p*-coumaric, *p*-hydroxybenzoic and protocatechuic acids, this pathway requires more study. Until more is known about the enzymes involved and, until all intermediates are identified, a true assessment of the relative importance of these pathways as compared with those suggested by the work of Crowden, is not possible.

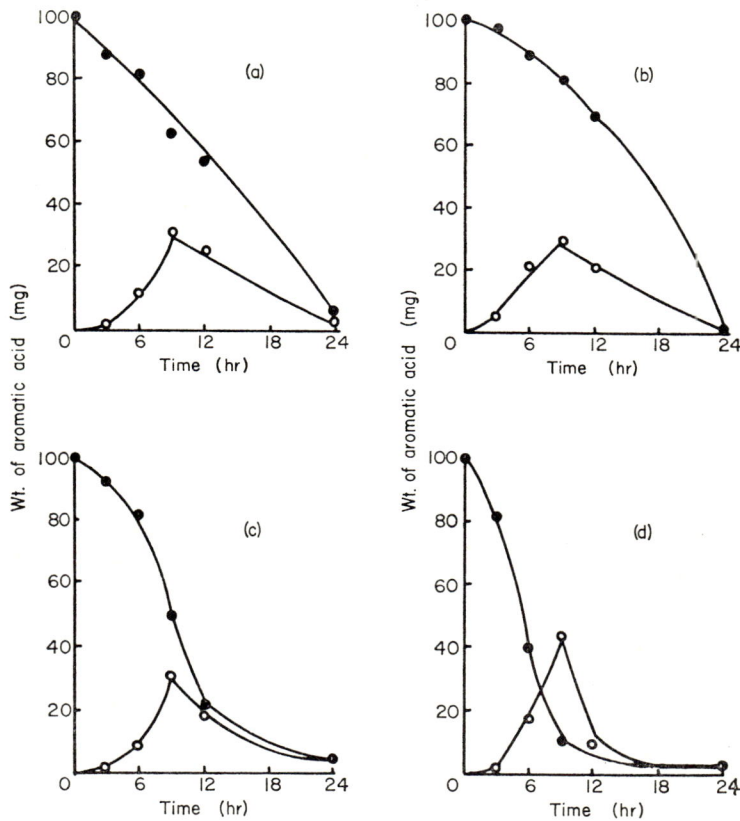

FIG. 6. Time-course of the disappearance of substrate (●) and formation of protocatechuic acid (○) in the presence of *Sporobolomyces roseus*. The substrates used were: (a) cinnamic acid; (b) benzoic acid; (c) *p*-hydroxybenzoic acid; (d) *p*-coumaric acid.

V. FORMATION OF MORE COMPLEX CINNAMOYL DERIVATIVES INCLUDING LIGNINS

Only two compounds, in which cinnamic acid or a hydroxylated derivative may be recognized as contributing to the overall structure, have been reported so far from Basidiomycetes. These are hispidin (XXIV), first isolated from *Polyporus hispidus* (Bu'Lock *et al.*, 1962; Bu'Lock and Smith, 1961; Edwards *et al.*, 1961) and later identified in *P. schweinitzii* (Ueno *et al.*, 1964) and cortisalin (XXV) isolated from *Corticium salicinum* (Gripenberg, 1952; Marshall and Whiting, 1957). Hispidin appears to be formed by extension of the side chain of cinnamic acid or a hydroxylated derivative by two equivalents of malonate. Cortisalin may be derived similarly from cinnamate and six equivalents of malonate.

Bu'Lock (1967) has given a preliminary account recently of his studies on hispidin biogenesis and the relation of this styrylpyrone to "lignification" in *Polyporus hispidus*. Hispidin is produced in sterile cultures of both *P. hispidus* and *P. schweinitzii* so that we have here an ideal system for studying a process very closely related to flavonoid biosynthesis in higher plants. Broadly speaking, the results so far indicate that hispidin synthesis is stimulated in cultures by the addition of phenylalanine, tyrosine and some of their derivatives including *p*-coumaric acid. Interestingly enough caffeic acid is not effective. Cinnamate metabolism in *P. schweinitzii* promises to be an especially interesting area for study since growth in this organism is reported to be promoted significantly by the addition of small amounts of ferulic acid to the culture medium (Robbins *et. al.*, 1963).

Hispidin
(XXIV)

Cortisalin
(XXV)

Bu'Lock (1967) believes that hispidin is oxidatively polymerized during the ripening of sporophores of *P. hispidus* and that the polymer becomes bound to cell material effecting chemical cross-linking. The product then serves as a "lignin" although it differs from lignins of higher plants in having no methoxyl content. Neither the "ethanol lignins" nor the sporophores of *Polyporus sulphureus*, *Fomes pinicola* or *Fomes applanatus*, three "bracket" fungi, yielded typical lignin products, e.g., vanillin and syringaldehyde, on alkaline nitrobenzene oxidation (Towers, 1951).

It would perhaps be better to await the results of further studies on Basidiomycetes before referring to the non-methoxylated, non-carbohydrate, structural component of the wall as "lignin" on the basis of its biological function. Studies by Lovyagina *et al.* (1960) and in Bu'Lock's laboratory on "chagi," a hot-water extract of *Poria obliqua*, reveal that a material much more like higher plant lignins is present. Degradation of this material yields *p*-hydroxyphenyl, guaiacyl and syringyl derivatives, compounds which are characteristic of the chemical degradation of plant lignins.

It is not likely that much headway can be expected in defining the exact structures of fungal lignins in the very near future. The lignins of higher plants have been studied for over a hundred years but many fundamental questions still remain unanswered. It would be a step forward, however, if systematic analyses of Basidiomycetes for polymers of the *p*-hydroxy-, guaiacyl- and syringyl-propane types could be carried out.

VI. COMPARISON OF CINNAMATE METABOLISM IN BASIDIOMYCETES AND IN HIGHER PLANTS

Table II represents an attempt to summarize, on the basis of all available information, what is known of cinnamate metabolism in Basidiomycetes in comparison with higher plants. In higher plants the cinnamate pathway leads to the production of lignins, flavonoids and related compounds as well as to benzyl derivatives. The significance of the production of benzoic acid derivatives *via* this route in higher plants is now known. We still do not know whether ring cleavage of any phenolic product of phenylalanine metabolism is achieved in higher plants.

In the few Basidiomycetes which have been studied, the production of cinnamic acid has been shown to lead through benzoic acid derivatives to CO_2. Flavonoids have not been discovered in these organisms but, in view of the occurrence of compounds like hispidin and cortisalin in some species, it would

TABLE II

Comparison of cinnamate metabolism in Basidiomycetes and higher plants

Metabolic feature	Higher plants	Basidiomycetes
Phenylalanine and tyrosine ammonia lyases	Ubiquitous	Absent in some
Hydroxylation, O-methylation esterification of cinnamate	+ Guaiacyl and syringyl derivatives typical	+
Formation of benzyl derivatives	+	–
Extension of side chain of cinnamate via polyketide pathway	+ Flavonoids typical	– Hispidin, cortisalin
Ring cleavage of phenyl- alanine derived metabolites	?	– Through protocatechuate
Lignin formation	+ Typical	?

not be extraordinary perhaps to find them here. Many fungi in this group are "woody" and it is likely that this woodiness is due in part to products of cinnamate metabolism. This remains to be proved.

What about other fungi or even bacteria? Very recently a species of *Streptomyces* has been found to accumulate cinnamide and to exhibit PAL activity

(L. C. Vining, personal communication).* It is likely that the synthesis of cinnamic acid and its derivatives is a basically important metabolic process common to groups of microorganisms other than the Basidiomycetes. This aspect of comparative phytochemistry should be a rewarding area for investigation in the coming decade.

REFERENCES

Bandoni, R. J., Moore, K., Subba Rao, P. V., and Towers, G. H. N. (1968). *Phytochemistry*, **7**, 205.
Bassett, E. W., and Tanenbaum, S. W. (1958). *Biochim. Biophys. Acta*, **28**, 247.
Birkinshaw, J. H., Chaplen, P., and Findlay, W. P. K. (1957). *Biochem. J.* **66**, 188.
Birkinshaw, J. H., and Findlay, W. P. K. (1940). *Biochem. J.* **34**, 82.
Bu'Lock, J. D. (1967). "Essays in Biosynthesis and Microbial Development", John Wiley & Sons, Inc. New York.
Bu'Lock, J. D., Leeming, P. R., and Smith, H. G. (1962). *J. chem. Soc.* 2085.
Bu'Lock, J. D., and Ryan, A. J. (1958). *Proc. chem. Soc.* **222**.
Bu'Lock, J. D., and Smith, H. G. (1961). *Experientia*, **17**, 553.
Crowden, R. K. (1967). *Can. J. Microbiol.* **13**, 181.
Edwards, R. L., and Elsworthy, G. C. (1967). *J. chem. Soc.* (c), 410.
Edwards, R. L., Lewis, D. G., and Wilson, D. V. (1961). *J. chem. Soc.* 4995.
Evans, W. C. (1963). *J. gen. Microbiol.* **32**, 177.
Fukuzumi, T. (1962). *Agr. Biol. Chem.* **26**, 447.
Gatenbeck, S., and Lonroth, I. (1962). *Acta Chem. Scand.*, **16**, 2298.
Gripenberg, J. (1952). *Acta Chem. Scand.* **6**, 580. **6**, 587.
Herr, R. R., Bergy, M. E., Eble, T. E., and Jahnke, J. K. (1960). Antimicrobial Agents Annual, p. 23.
Ichihara, K., Ikeda, S., and Sakamoto, Y. (1956). *J. Biochem.* Tokyo, **43**, 129.
Jones, J. D., Smith, B. S. W., and Evans, W. C. (1952). *Biochem. J.* **51**, xi.
Kluyver, A. J., and Van Zijp, J. C. M. (1951). *Leewenhoek ned. Tijdschr.* **17**, 315.
Koukol, J., and Conn, E. E. (1961). *J. biol. Chem.* **236**, 2692.
Kubota, T., and Naya, K. (1954). *Chem. Ind.* 1427.
List, P. H., and Freund, B. (1966). *Naturwiss.* **53**, 585.
Lovyagina, E. V., Shivrina, A. N., and Platonova, E. G. (1960). *Biokhimya*, **25**, 640.
Marshall, D., and Whiting, M. C. (1957). *J. chem. Soc.* 537.
Meister, A. (1965). "Biochemistry of the amino acids", Vol. II, p. 884, Academic Press, New York.
Minamikawa, T., and Uritani, I. (1965). *J. Biochem. (Japan)*, **57**, 678.
Moore, K., Subba Rao, P. V., and Towers, G. H. N. (1967). *Life Sciences*, **6**, 2629.
Moore, K., Subba Rao, P. V., and Towers, G. H. N. (1968). *Biochem. J.* **106**, 507.
Moore, K., and Towers, G. H. N. (1967). *Can. J. Biochem.* **45**, 1659.
Namura *et al.*, *J. Antibiotics* 20A, **55**, 1967.
Nord, F. F., and Vitucci, J. C. (1947). *Arch. Biochem.* **14**, 243.
Ogata, K., Uchiyama, K., and Yamada, H. (1966). *Agr. Biol. Chem.* **30**, 311.
Ogata, K., Uchiyama, K., and Yamada, H. (1967a). *Agr. Biol. Chem.* **31**, 200.
Ogata, K., Uchiyama, K., Yamada, H., and Tochikura, T. (1967b), *Agr. Biol. Chem.* **31**, 600.
Paris, R., Durand, M., and Bonnet, J. L. (1960). *Ann. pharm. franc.* **18**, 769.

---

* Dr. L. Vining of the National Research Council has drawn my attention to the presence of cinnamide in cultures of *Streptomyces caespitosus*, *S. reticuli* and *S. verticillatus*. (Herr *et al.*, 1960; Namura *et al.*, 1967; Wakiki *et al.*, 1958).

Power, D. M., Towers, G. H. N., and Neish, A. C. (1965). *Can. J. Biochem.* **43**, 1397.
Robbins, W. J., Hervey, A., Page, A. C., Gale, P. H., Hoffman, C. H., Moscatelli, E. A., Koninszy, F. R., Smity, M. C., and Folkers, K. (1963). *Mycologia*, **55**, 742.
Rogoff, M. H. (1961). *Adv. appl. Microbiol.* **3**, 193.
Sakamoto, Y., Mitsuhashi, T., and Ichihara, K. (1958). *J. Biochem. Tokyo* **45**, 1.
Schubert, W. J. (1965). "Lignin Biochemistry", Academic Press, New York.
Shimazono, H. (1959). *Arch. Biochem. Biophys.* **83**, 206.
Shimazono, H., Schubert, W. J., and Nord, F. F. (1958). *J. Am. chem. Soc.* **80**, 1992.
Subba Rao, P. V., Moore, K., and Towers, G. H. N. (1967). *Can. J. Biochem.* **45**, 1863.
Towers, G. H. N. (1951). M.Sc. Thesis, McGill University, Montreal, Canada.
Trecanni, V. (1963). *Progr. industr. Microbiol.* **4**, 1.
Ueno, A., Fukushima, S., Saiki, Y., Harada, T. (1964). *Chem. pharm. Bull.* **12**, 376.
Van Sumere, C. F., van Sumere-de Preter, C., Vining, L. C., and Ledingham, G. A. (1957). *Can. J. Microbiol.* **3**, 847.
Wakaki, S., Marumo, H., Tomioka, K., Shimizu, G., Kato, E., Kamada, H., Kudo, S., and Fujimoto, Y. (1958). *Antibiotics and Chemotherapy* **8**, 228.
Young, M. R., Towers, G. H. N., and Neish, A. C. (1966). *Can. J. Bot.* **44**, 341.

CHAPTER 10

Flavonoids and Photomorphogenesis in Peas

ARTHUR W. GALSTON

Department of Biology, Yale University, New Haven, Connecticut, U.S.A.

I.	Introduction	193
II.	Growth Responses of the Pea to Light	194
III.	Identification of Flavonoids Affecting IAA Oxidase Activity	196
IV.	Relation of Flavonoid Content to Growth	197
V.	Effect of Light on Uptake of Flavonoid Precursors	198
VI.	Flavonoids in Pea Tendrils	200
VII.	Other Relevant Recent Investigations	203
VIII.	Conclusion	203
	References	204

I. INTRODUCTION

The etiolated pea seedling possesses a long, slender, virtually unpigmented stem with very long internodes, reduced scale-like unpigmented leaves and a recurved apical hook immediately below the terminal bud. A plant of similar age grown in the light has short, thick, pigmented stems with no apical hook and well-expanded green leaves. Light may therefore be said to produce a minimum of four effects in this seedling: inhibition of stem elongation, promotion of leaf growth, promotion of pigment synthesis and uncoiling of the apical hook.

For almost twenty years, I have been involved in an attempt to understand some of the biochemistry involved in producing these growth alterations following absorption of the effective quanta. Among the early biochemical consequences of irradiation are profound alterations in the quantity and quality of flavonoids produced in the various organs. Because it appears possible that such alterations in flavonoids may be meaningfully related to the alteration of growth patterns, I shall attempt to sketch the history and our present understanding of flavonoids and photomorphogenesis in peas.

The first thing to note is that three of the four responses noted are under the control of phytochrome (Parker *et al.*, 1949); the fourth, light promotion of greening, is more complex, being partly under the control of phytochrome and partly the result of light absorption by protochlorophyll and carotenoids. For simplicity, I shall omit this last category and concentrate on the three growth responses.

II. GROWTH RESPONSES OF THE PEA TO LIGHT

While studying the growth of excised cylinders of epicotyl originating from just under the apical buds of seven-day old seedlings, we found (Galston and Baker, 1953) that red light depressed endogenous growth and also increased, by about two orders of magnitude, the concentration of the auxin indoleacetic acid (IAA) required to elicit optimal growth (Fig. 1). A single intense flash of red light, delivering *ca.* 2 kiloergs/cm² to the plant surface, produced the

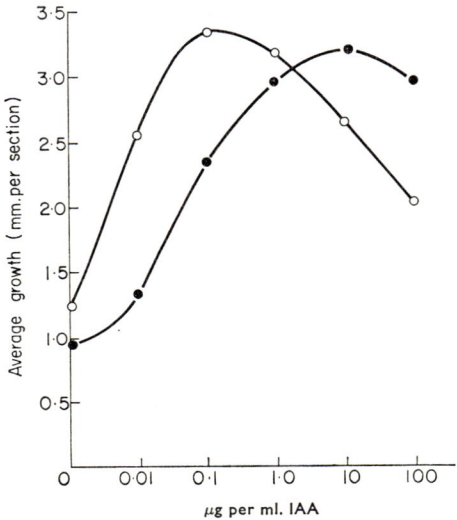

FIG. 1. Decreased sensitivity to auxin induced by preirradiation with red light. ○ dark-grown epicotyls; ● red-grown epicotyls.

maximal effect if given 18 h before excision of the sections for the growth test, which was itself conducted in complete darkness. If the red light was given 6–8 h before the growth test, it produced a half-maximal effect; if delivered 24 h before the growth test, it was virtually ineffectual. From these observations we concluded that light initiated certain biochemical changes which grew in physiological effectiveness up to 18 h, then rapidly decayed to zero at 24 h. Red light also decreased the amount of IAA which disappears from the solution bathing the epicotyl sections, through a diminution of either auxin uptake or auxin destruction *in vivo* or in the ambient solution (Table I).

Later, Hillman and I (1957) found that inductive red light treatments, similar to those which affected epicotyl growth, lowered IAA oxidase activity of a homogenate of the treated tissue. This effect of red is reversed by far-red light (Table II), and shows the same kinetics as the growth effect. In bud tissue, the diminished IAA oxidase activity is due to increased production of a dialyzable inhibitor. If the buds are excised prior to irradiation, neither

TABLE I

The effect of light pretreatment on the removal from solution of exogenously supplied IAA. Twelve epicotyl sections immersed in 5 ml of solution containing $10 \mu g/ml$ IAA

Experiment no.	$\mu g/ml$ IAA removed from solution in 16 hr[a]	
	D Series	R Series
1	3·4	2·3
2	6·6	4·0
3	5·5	1·7
4	7·4	4·4
5	5·9	4·1

[a] D series; pea seedlings grown in darkness, R series; pea seedlings grown in continuous red light.

TABLE II

IAA Oxidase activity of pea buds 16 hours after exposure of intact plants to red and far-red (FR) radiation in sequence[a]

Treatment	IAA oxidase activity[b]	Inhibition, %
Dark controls	8·3	(0)
FR only	8·0	4
Red only	0·7	92
Red-FR	7·4	11
Red-FR-Red	1·0	88
Red-FR-Red-FR	7·6	8

[a] Red light exposure 2 min, total energy 30 kiloergs/cm^2. FR exposure 2 minutes (higher intensity. Less than 1 min elapsed between successive exposures.
[b] Activity measured as $\mu gm/ml$ IAA destroyed in 20 minutes. Each 10-ml reaction mixture contained 35 $\mu gm/ml$ IAA at the start, 10^{-4} M 2,4—dichlorophenol, and bud extract equivalent to 24 mg fresh wt.

growth, IAA oxidase activity, nor dialyzable inhibitor content is altered. Attempts to substitute for the rest of the plant by adding materials to the media bathing the excised buds were unsuccessful. In our view, this work heightened the possibility that the phytochrome-linked enhancement of the synthesis of an IAA oxidase inhibitor was causally related to the control of growth by light. An attempt was therefore initiated to isolate and identify the inhibitor.

III. IDENTIFICATION OF FLAVONOIDS AFFECTING IAA OXIDASE ACTIVITY

Two groups, working independently, identified flavonoid conjugates in the inhibitory fraction (Mumford *et al.*, 1961; Furuya *et al.*, 1962). Furuya and I (1965) found etiolated peas to contain kaempferol-3-triglucoside (I)* (KG)

Kaempferol 3-triglucoside (I)*

Kaempferol 3-(*p*-coumaroyltriglucoside) (II)

Quercetin 3-triglucoside (III)

Quercetin 3-(*p*-coumaroyltriglucoside) (IV)

and its *p*-coumaric acid ester (II) KGC), while green peas contained in addition the analogous quercetin compounds (III) and (IV) (QG and QGC). Mumford and colleagues (1961) believed that the inhibitor was a kaempferol derivative and that this was a hexaglucoside; Furuya and I believed that both kaempferol

* Glc = glucose; linkage in trisaccharide probably $\beta,1\rightarrow2$; position of acyl linkage in II and IV still undetermined.

conjugates were promoters of IAA oxidase activity at physiological concentrations, and that the quercetin derivatives were the inhibitors. This is in accord with earlier findings (Goldacre *et al.*, 1953) showing that monophenols tend to promote and *o*- and *p*-diphenols tend to inhibit IAA oxidase activity.

IV. RELATION OF FLAVONOID CONTENT TO GROWTH

Using the paper chromatographic technique for separating the flavonoids developed by Furuya (1962), Furuya and Thomas (1964) investigated the relation between enhanced growth of plumules following irradiation and their

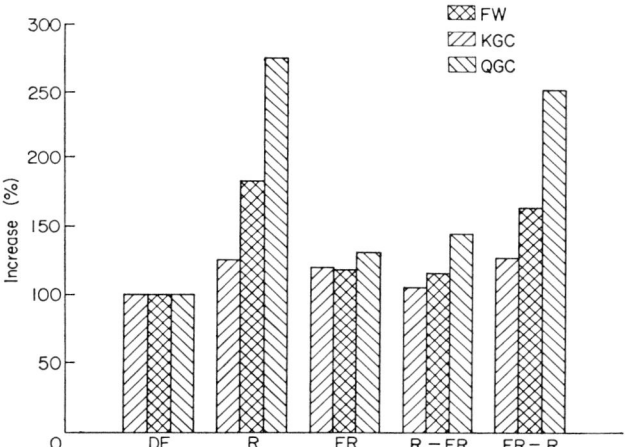

FIG. 2. Variations in KGC and QGC content and of growth of dark grown peas after irradiation with combinations of red (R) and far-red (FR) light. Flavonoid content and growth are expressed as a percentage of dark control (DF). Red light exposure; 546 ergs/cm^2/sec for 3 min (98 k ergs/cm^2). Far-red light exposure: 1173 ergs/cm^2 sec for 12 min (845 k ergs/cm^2). Fresh weight (FW) was measured as the average weight of the terminal buds. KGC and QGC contents were measured as μmoles/g fresh weight.

content of flavonoids. They found KGC content to rise within eight hours, about as rapidly as increased plumule weight could be detected; KG level was unaffected by light. Because the photoeffect on growth was saturated at *ca.* 0·2 k ergs/cm^2, while 2·0 k ergs/cm^2 were required to saturate the enhanced pigment synthesis effect, they reasoned that light enhances growth and pigment synthesis independently, and that no causal relation was implied by their data. Reinvestigating this problem, Bottomley *et al.* (1965, 1966) found that the paper-chromatographic technique of Furuya did not separate KGC from QGC; on a silica-gel column developed with a butanol-boric acid mixture, these compounds are well resolved. Using this improved analytical technique, Bottomley *et al.* were able to show that red light affected QGC, not KGC (Fig. 2). Thus, one effect of phytochrome activation appeared to be a control of the hydroxylation pattern of flavonoids of the plumule. In the internode tissue, on the other hand, photo-conversion of phytochrome by red light caused an increase in KGC and especially in KG (Fig. 3); quercetin conjugates could not be detected (Russell and Galston, 1967). Thus, the exact chemical conse-quences of phytochrome conversion depend on the nature of the tissue. What happens in the pea seems to make sense for the control of growth: red light promotes the growth of buds, and simultaneously promotes the synthesis of

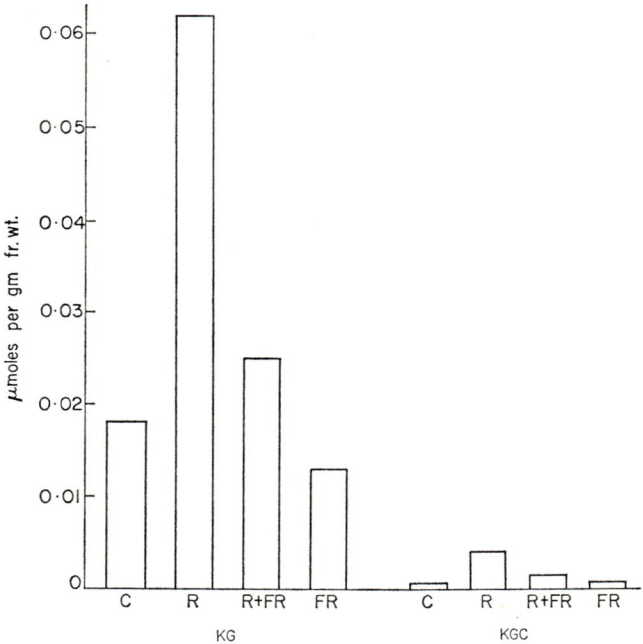

FIG. 3. The effects of red and far-red light on the levels of KG and KGC in internode tissues derived from 5 mm subapical segments of intact etiolated progress seedlings. The segments were marked off, treated at zero time and harvested 24 hr later. C = control; R = red; FR = far red.

the auxin-sparing substance, quercetin. Contrariwise, in the stem, where phyto-chrome conversion leads to growth inhibition, the synthesis of kaempferol, a cofactor for auxin destruction, is promoted. This would not explain, however, why red light caused a diminution in the disappearance of IAA from a solution surrounding elongating epicotyl segments.

V. EFFECT OF LIGHT ON UPTAKE OF FLAVONOID PRECURSORS

Since phytochrome obviously controls flavonoid synthesis in peas, it seemed desirable to investigate the chemical locus at which control is exerted. This was attempted by Goren and Galston (1966) who fed labelled substrates to etiolated epicotyls excised at the cotyledonary level and studied their uptake and incorporation with and without phytochrome conversion. Both bud growth and ^{14}C-sucrose appearance in the bud were promoted by red light, the latter effect being quantitatively the more important (Fig. 4). In stems, both quantities were depressed, and the effect on ^{14}C-sucrose was again the greater. Because the sucrose uptake effect was more rapid, of greater magnitude and saturated

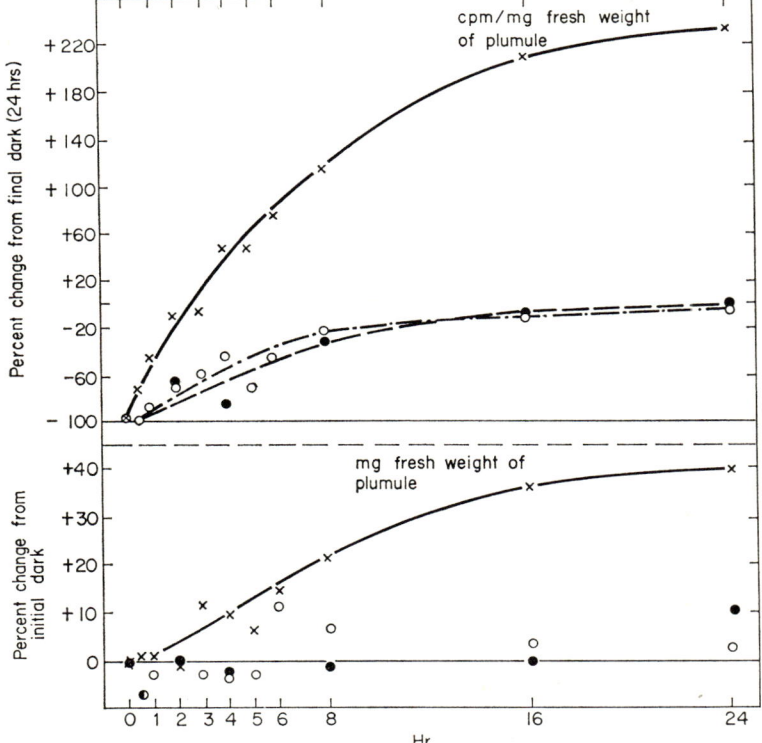

FIG. 4. Kinetics of ^{14}C-sucrose incorporation and of plumule growth. × red light; ○ far-red light; ● dark control.

by lower energies than the effect on growth, its causal relation to growth regulation was hypothesized.

There are several obscure aspects of this reaction: buds devoid of stem will show neither enhanced growth nor enhanced ^{14}C-sucrose uptake after red irradiation, and the longer the length of epicotyl to which they are attached, the greater is the promotive effect of red light. This suggests control (restraint) of the rate of sucrose passage into the bud, and its partial release by phytochrome conversion. The effect is most dramatic with sucrose; smaller red-light effects are shown with labelled fructose, glucose, maltose and ribose in that order;

various amino acids and organic acids showed poor uptake and relatively small red light effects. Gibberellic acid applied up to two hours after red light inhibited the light effect; auxins, cytokinins, and inhibitors of gibberellin biosynthesis were without effect (Goren and Galston, 1967). Hanada and Galston (unpublished results) isolated kaempferol and quercetin conjugates from dark-grown or red-irradiated buds fed ^{14}C-sucrose in the Goren-Galston manner. Paradoxically, after irradiation the specific activity of KG + KGC (which do not increase in quantity) increased much more than the specific activity of QG + QGC (which do increase in quantity) (Table III). Because of this, it appeared possible that light actually promotes the synthesis of kaempferol derivatives a little, and that quercetin derivatives are formed directly by

Table III

The effect of red light on the radioactivity of kaempferol and quercetin complexes extracted from completely etiolated or red-irradiated pea epicotyls supplied basally with U-^{14}C-sucrose

| | cpm/μM flavonoid | | | |
| | KG and KGC | | QG and QGC | |
Expt	Dark	Red	Dark	Red
1	0·98	5·48	2·18	5·73
	1·91	6·97	5·97	7·47
2	1·33	4·40	11·39	9·97
	1·52	4·37	8·99	9·09
Avg.	1·44	5·31	7·13	8·07

Flavonoids extracted 24 hrs after irradiation.

hydroxylation of kaempferol compounds once the latter have surpassed a critical level. Further investigations along these lines are proceeding in several laboratories.

VI. FLAVONOIDS IN PEA TENDRILS

In these investigations, as in most others in biology, many serendipitous discoveries appear, and attempt to seduce the investigator away from his original line of investigation. On occasion, we have succumbed to the lure of such divertissements, sometimes with great pleasure and reward. After having noted that pea leaf tendrils are unusually rich in flavonoids, mainly of the Q-type (Bottomley et al., 1966), we investigated the physiology of these rapidly moving organs (Jaffe and Galston, 1966a, 1966b, 1967a, 1967b). The initial response to touch of the ventral half of the tendril is a contraction, followed by unequal expansion of dorsal and ventral halves. The contraction is associated with a loss of water from the ventral cells, presumably due to a sudden change of permeability in the membranes of those cells. We found such changes to be correlated with ATPase activity, in that ATP in the tendril declined and inorganic phosphate increased following stimulation. It is of interest that flavonoids, which are most abundant near the sensitive tip of the tendril (Table IV), decline rapidly following stimulation. The decline is due entirely to QGC (Table Va and b); furthermore, if physiological levels of quercetin are applied to tendrils *in vitro*, their response to mechanical stimuli is inhibited (Table VI). From this, one might infer that flavonoid (quercetin) breakdown is requisite for the APTase activation. This hypothesis, which is supported by the comparative kinetics of curvature and QGC breakdown, would assign to flavonoids a broadly significant regulatory role in contractile and membranous phenomena.

TABLE IV

Topographic distribution of flavonoids in unstimulated pea tendrils from plants grown in continuous light

Part of tendril	Total flavonoids
	μmoles per g fr wt
Apical fourth	7.21 ± 0.07^a
Second fourth	5.62 ± 0.81
Third fourth	3.10 ± 0.20
Basal fourth	1.86 ± 0.07

[a] Standard error.

Each sample was extracted from about 10 mg of tissue. Data derived from three replicate experiments.

The diversion into studies of tendril physiology whetted our interest in rapid plant movements, which surprisingly led back to phytochrome. Fondéville *et al.* (1966, 1967) had reported that the nyctinastic closure of *Mimosa* leaf pinnules and pinnae was under the control of phytochrome, and that closure was rapid only when the last light prior to the dark period is red, not far-red. We were able to duplicate this in the related, but non-thigmotropic genus *Albizzia* (Jaffe and Galston, 1967c). We also showed that along with movement, there was a rapid efflux of conductometrically-measurable materials, in part attributable to H^+. This apparent phytochrome control of membrane permeability is not sensitive to actinomycin D, and so presumably does not depend on the synthesis of m-RNA or proteins (although a long-lived messenger might becloud this conclusion). It then occurred to us that this apparent effect of phytochrome on membrane permeability might satisfactorily explain the Goren-Galston data, in which movement of labelled sucrose up the epicotyl and into the bud of an excised etiolated pea is enhanced by

TABLE Va

Flavonoid contents of tendrils *in situ* on plants grown in continuous light

	μmoles per g fr wr	
Condition of tendrils	KGC	QGC
Not coiled	$0·21 \pm 0·12$	$2·76 \pm 0·51$
Coiled	$0·30 \pm 0·20$	$1·15 \pm 0·40$

Samples were taken of tendrils which had either not yet coiled or had wound firmly around supports. Each datum represents the mean of 4 experiments and is followed by its standard error.

Table Vb

Correlation of kinetics of coiling with decrease in QGC in tendrils allowed to coil *in situ* in continuous light

Min after stimulation	Curvature	Total flavonoids
	degrees	μmoles/g fr wt
0	248	2·32
20	429	2·00
40	573	1·43
60	709	1·41
90	677	0·94
120	663	0·65
Correlation coefficient (5)		$-0·874^a$

[a] Significant at the 5% level.

TABLE VI

The effect of aqueous extracts of tendrils and of QGC on the coiling of excised tendrils in the light

Addendum	Net Degrees Curvature in 20 hr
None	852 ± 156^a
Aqueous extract of unstimulated tendrils	548 ± 57
Aqueous extract of coiled tendrils	674 ± 144
None	720 ± 60
QGC, 10 μM	488 ± 100

[a] Standard error.

The aqueous extracts were made from excised tendrils which were either unstimulated (*ca* 75° curvature) or shaken and allowed to coil for 2 hours (*ca* 522° curvature).

phytochrome activation. Whether flavonoids participate in this type of control remains for future investigations to decide.

VII. OTHER RELEVANT RECENT INVESTIGATIONS

One enzyme involved in the biosynthesis of flavonoids is L-phenylalanine ammonia lyase, which converts L-phenylalanine to *trans*-cinnamic acid (Koukol and Conn, 1961). Many groups of workers have shown that light increases the activity of this enzyme, but recently, two independent investigations have shown strict phytochrome control of the synthesis of this enzyme in peas (Attridge and Smith, 1967; Russell and Conn, unpublished results). An increase in the hydroxylation level of phenolic acids (Engelsma, 1965) and lignin precursors (Stafford, 1965) following phytochrome photoactivation has also been reported, amplifying the possibility of control of hydroxylation by phytochrome. In view of the demonstration of the rapid control of membrane permeability by phytochrome under conditions where protein synthesis probably does not occur, these slower effects, involving the synthesis of proteins, must probably be regarded as secondary. The recent demonstration by Tanada (1968) of a phytochrome-mediated rapid alteration of the adhesion of barley root segments to Pyrex glass is best interpreted in terms of direct alteration of surface (membrane ?) characteristics, perhaps by the efflux of positively charged (H^+?) materials.

VIII. CONCLUSION

The activation of phytochrome by red light in etiolated peas leads to several growth effects (inhibition of stem elongation, promotion of leaf growth, opening of the apical hook) which are accompanied by the increased synthesis of flavonoids, as well as enzymes involved in flavonoid biogenesis. In some instances, the qualitative or quantitative alteration of flavonoid patterns occurs rapidly enough to be causally related to the growth changes. If flavonoids do control growth rate or patterns, this may occur by virtue of their effect on indoleacetic acid (IAA) destruction. Thus, the decreased growth of internodes may result from the increased synthesis of kaempferol conjugates, which act as cofactors for IAA oxidation, while the increased growth of leaves may result from the increased synthesis of quercetin conjugates, which inhibit IAA destruction. Flavonoids may, however, intervene in growth in other ways as well.

Pea tendrils contain large quantities of flavonoids, especially quercetin conjugates. When the tendril is mechanically stimulated, the level of quercetin declines rapidly as curvature proceeds, and the addition of exogenous quercetin retards curvature. This may be related to effects on an ATPase which is activated following mechanical stimulation.

In the pea plant, therefore, flavonoids are not inert end products of secondary plant metabolism, nor are they likely to be without some function in growth regulation.

REFERENCES

Attridge, T. H., and Smith, H. (1967). *Biochim. Biohys. Acta*, **148**, 805.
Bottomley, W., Smith, H., and Galston, A. W. (1965). *Nature, Lond.* **207**, 1311.
Bottomley, W., Smith, H., and Galston, A. W. (1966). *Phytochemistry*, **5**, 117.
Engelsma, G. (1965). *Nature, Lond.* **208**, 1117.
Fondéville, J. C., Borthwick, H. A., and Hendricks, S. B. (1966). *Planta*, **69**, 357.
Fondéville, J. C., Schneider, M. J., Borthwick, H. A., and Hendricks, S. B. (1967). *Planta*, **75**, 228.
Furuya, M. (1962). Ph.D. Thesis, Yale University.
Furuya, M., and Galston, A. W. (1965). *Phytochemistry*, **4**, 285.
Furuya, M., Galston, A. W., and Stowe, B. B. (1962). *Nature*, **193**, 456.
Furuya, M., and Thomas, R. G. (1964). *Pl. Physiol.* **39**, 634.
Galston, A. W., and Baker, R. S. (1953). *Am. J. Bot.* **40**, 512.
Goldacre, P. L., Galston, A. W., and Weintraub, R. L. (1953). *Arch. Biochem. Biophys.* **43**, 358.
Goren, R., and Galston, A. W. (1966). *Pl. Physiol.* **41**, 1055.
Goren, R., and Galston, A. W. (1967). *Pl. Physiol.* **42**, 1087.
Hillman, W. S., and Galston, A. W. (1957). *Pl. Physiol.* **32**, 129.
Jaffe, M. J., and Galston, A. W. (1966a). *Pl. Physiol.* **41**, 1014.
Jaffe, M. J., and Galston, A. W. (1966b). *Pl. Physiol.* **41**, 1152.
Jaffe, M. J., and Galston, A. W. (1967a). *Pl. Physiol.* **42**, 845.
Jaffe, M. J., and Galston, A. W. (1967b) *Pl. Physiol.* **42**, 848.
Jaffe, M. J., and Galston, A. W. (1967c). *Planta*, **77**, 135.
Koukol, J. E., and Conn, E. E. (1961). *J. biol. Chem.* **236**, 2692.
Mumford, F. E., Smith, D. H., and Castle, J. E. (1961). *Pl. Physiol.* **36**, 752.
Parker, M. W., Hendricks, S. B., Borthwick, H. A., and Went, F. W. (1949). *Am. J. Bot.* **36**, 194.
Russell, D. W., and Galston, A. W. (1967). *Phytochemistry*, **6**, 791.
Stafford, H. A. (1965). *Pl. Physiol.* **40**, 130.
Tanada, T. (1968). *Proc. natn. Acad. Sci.* (U.S.) **59**, 376.

Author Index

Numbers in italics are those pages on which References are listed.

A

Abe, A., 144, 151, 161, *163*
Abelson, P. H., 108, *119*
Abrol Y. P., 71, *72*
Aebi, A., 149, 150, 162, *163*
Agranoff, B. W., 751, *89*
Ahmad, A., 53, 65, 66, *72, 74*
Alihaud, G. P., 92, *106*
Airy Shaw, H. K., 124, *136*
Akazawa, T., 49, 51, *72, 73*
Albers, H., 49, *72*
Alberts, A. W., 91, 92, *105, 106*
Allen, J., 70, *72*
Alston, R. E., 109, *119*
Anderson, D. G., 75, 78, 84, *90*
Anderson, L., 49, *73*
Anderson, L. A. P., 144, 149, *163*
Andreae, W. A., 122, *136*
Archer, B. L., 78, *89*
Archie, W. C., 116, *120*
Arndt, R. R., 111, *119*
Attridge, T. H., 203, *204*
Augier, J., 154, *163*
Aynechi, L., 141, *164*

B

Bachmann, G., 152, *164*
Bailey, I. H., 152, *163*
Bailey, R. W., 49, *73*
Baker, R. S., 194, *204*
Balocchi, M., 131, *138*
Bandoni, R. J., 181, *190*
Barnard, D., 78, *89*
Barreira, J. B., 147, *163*
Barron, E. J., 91, *105*
Bartels, C, T., 99, *105*
Barton, D. H. R., 113, *119*, 142, 147, *163*
Bassett, E. W., 184, *190*
Bates, R. B., 149, *164*

8

Bate-Smith, E. C., 167, 173, 175, *177*
Batterham, T. S., 31, *44*
Battersby, A. R., 113, 114, *119*
Bawdekar, A. S., 155, *164*
Becker, W., 50, *72*
Bell, E. A., 67, *72, 73*
Benn, M. H., 63, *72*
Benthin, U., 50, *72*
Ben-Yehoshua, S., 53, *72*
Bergy, M. E., 189, *190*
Bezard, J., 100, *105*
Bhat, H. B., 118, *120*
Bhattacharyya, S. C., 142, 152, 155, 156, *164, 165*
Birkinshaw, J. H., 180, *190*
Bleichert, E. F., 55, 57, *72*
Bloch, K., 91, 93, 94, 95, 96, 98, 99, 100, 104, *105, 106*
Bloomfield, D. K., 93, 95, *105*
Blumenthal-Goldschmidt, S., 60, 61, 66, 67, *72*
Bohlmann, F., 110, *119*, 126, *136*
Bonnet, J. L., 180, *190*
Bornfleth, H., 122, *137*
Borthwick, H. A., 193, 201, *204*
Bottomley, W., 197, 200, *204*
Bourdu, R., 133, *136*
Bove, J. M., 93, *106*
Breton, J. R., 147, *163*
Breuer, J., 101, *106*
Brink, C. M., 144, 149, *163*
Britton, G., 79, 80, 82, 84, *90*
Brooks, J. L., 92, 93, *105*
Brown, S. H., 111, *119*
Bruns, F. H., 52, *73*
Brysk, M. M., 70, *73*
Büchi, J., 149, 150, 162, *163*
Bu'Lock, J. D., 104, *105*, 180, 184, 187, 188, *190*
Burlingame, A. L., 141, *164*
Burnett, A. R., 117, *119*
Butler, B. G., 53, *73*

Butler, G. W., 49, 50, 53, 54, 55, 57, 58, 59, 60, 61, 63, 64, 65, 66, 67, *72*, *73*, *74*

C

Canvin, D. T., 102, 104, *105*
Carrazsoni, E. P., 111, *119*
Cartier, D., 133, *136*
Caspi, E., *89*
Cassady, Y. M., 141, *164*
Castle, J. E., 196, *204*
Cavallito, Ch. I., 152, *163*
Chalk, L., 125, 127, *137*
Chaplen, P., 180, *190*
Charlton, J. M., 82, 84, *89*, *90*
Chenery, E. M., 122, 123, *136*
Chikamatsu, H., 142, 158, *163*, *164*
Chisholm, M. D., 63, 64, *74*
Clayton, R. B., 87, *89*, *90*
Coburn, R. A., 48, *73*
Cockbain, E. G., 78, *89*
Conn, E. E., 49, 50, 51, 52, 53, 54, 55, 56, 57, 58, 59, 60, 61, 63, 64, 66, 67, 68, 71, *72*, *73*, *74*, 182, *190*, 203, *204*
Cooper, J. M., 71, *73*
Corey, E. J., 87, *89*
Cornforth, J. W., 75, 77, 78, 82, 87, *89*, *90*
Cornforth, R. H., 77, 78, 82, 87, *89*
Corpe, W. A., 70, *73*
Cranmer, M. F., 31, *45*, 122, 123, *136*
Criddle, R. S., 49, *74*
Cronquist, A., 124, 127, *136*, 139, 154, *163*
Crowden, R. K., 183, 184, 185, *190*

D

Debuch, H., 99, *105*
Deenen, L., van, 199, *105*
Deuel, P. G., 142, *163*
Dilleman, G., 47, 48, *73*
Djerassi, C., 111, *119*, 150, *164*
Dobson, T. A., 113, *119*
Dolejš, L., 142, 146, 147, 149, *163*, *164*, *165*
Donninger, C., 77, 78, 82, 87, *89*
Dreiding, A. S., 111, *120*
Drozdz, B., 152, 153, *163*
Duarte, A. P., 111, *119*
Dunnill, P. M., 67, 68, *73*
Dunstan, W. R., 71, *73*

Durand, M., 180, *190*
Dyke, S. F., 22, *44*

E

Eade, R. A., 41, 44
Eames, A. J., 125, *136*
Eberhardt, F. M., 93, *105*
Eble, T. E., 189, *190*
Edwards, R. L., 180, 187, *190*
Eggerer, H., 75, *89*
Ehrendorfer, R., 136, *136*
Elsworthy, G. C., 180, *190*
Emberger, L., 124, *136*
Emery, T. F., 66, *74*
Engelsma, G., 203, *204*
Engler, A., 124, *136*
Erwin, J., 95, 100, *105*
Eschenhif, E., 50, *72*
Ettlinger, M., 110, *119*
Evans, W. C., 183, *190*
Exell, A. W., 123, *136*
Eyde, R. H., 127, 131, *136*

F

Fajards, M., 147, *163*
Farreira, J. M., 111, *119*
Farrier, D. S., 116, *120*
Fatianoff, O., 109, *120*
Faure, A., *127*, *136*
Favre-Bonvin, J., 53, 57, *73*
Fawcett, C. H., 122, *136*
Fiedler, L., 52, *73*
Findlay, W. P. K., 180, *190*
Finnemore, H., 71, *73*
Flores, S. E., 111, *119*
Floss, H. G., 68, *73*
Folkers, K., 188, *191*
Fondéville, J. C., 201, *204*
Forchielli, E., *89*
Fowden, L., 67, 68, 72, *73*
Fraenkel, G. S., 72, *73*
Freund, B., 180, *190*
Frohne, D., 134, *137*
Fujimoto, Y., 189, *191*
Fukushima, S., 180, 187, *191*
Fukuzumi, T., 184, *190*
Fulco, A. J., 98, *105*
Furuya, M., 196, 197, *204*

G

Gale, P. H., 188, *191*
Galliard, T., 102, *105*

Galston, A. W., 194, 196, 197, 198, 199, 200, 201, *204*
Gander, J. E., 51, 52, 63, *73*
Gatenbeck, S., 184, *190*
Geiger, U. P., 11, *119*
Geissman, T. A., 17, 20, *44*, *45*, 49, *73*, 142, 156, 158, *163*
Gentili, B., 41, *45*
Gilberg, F. B., 111, *119*
Gilham, P. T., 142, *163*
Giovanelli, J., 69, *73*
Giza, Y-H., 67, *74*
Goad, L. J., 87, *89*, *90*, 116, *119*
Goffeau, A., 93, *106*
Goldacre, P. L., 197, *204*
Goldfine, H., 94, *105*
Goldman, P., 91, *105*
Goldschmidt, O., 112, *120*
Gonsáles, A. G., 147, *163*
Goodwin, H., 122, *137*
Goodwin, T. W., 75, 78, 79, 80, 82, 84, 87, 89, *89*, *90*
Goren, R., 199, 200, *204*
Gorenflot, R., 133, *136*
Gourlay, H. W., 125, *137*
Govindachari, T. R., 141, *163*
Gozmán, A., 150, 160, *165*
Gripenberg, J., 180, 188, *190*
Grouiller, A., 31, 32, *45*
Gundersen, A., 124, *137*
Gupta, R. N., 116, *119*

H

Hadaway, M. C., 102, *105*
Hadwiger, L., 68, *73*
Hahlbrock, K., 58, 59, 60, 63, *73*
Haigh, W. G., 104, *105*
Hall, S. W., 98, 101, 102, *106*
Hallier, H., 134, *137*
Hankes, L. V., 70, *73*
Hansel, R., 17, *45*
Harada, T., 180, 187, *191*
Harborne, J. B., 17, 20, *44*, *45*, 109, *119*, 172, *177*
Harmatha, J., 150, 160, 161, *163*
Harris, P., 95, 99, 100, 104, *105*, *106*
Harris, R. V., 95, 100, *105*
Haslewood, G. A. D., 108, 112, *120*
Haverkate, F., 99, *105*
Hayashi, M., 144, 151, *164*
Hegnauer, R., 47, *73*, 109, *120*, 121, 122, 127, 133, 134, *137*, 140, 144, 146, *163*, 168, 169, *177*

Helferich, B., 49, *73*
Hendricks, S. B., 193, 201, *204*
Hendrickson, H. R., 68, *73*
Hendrickson, J. B., 142, *163*
Henning, V., 75, *89*
Henry, T. A., 71, *73*
Herout, V., 140, 141, 142, 146, 147, 149, 150, 151, 152, 153, 154, 155, 158, 160, 161, 162, *163*, *164*, *165*
Herr, R. R., 189, *190*
Hervey, A., 188, *191*
Herz, W., 140, 142, 144, 147, 156, 158, *163*, *164*, *165*
Heskett, M. G., 66, *73*
Highet, R. J., 31, *44*
Hikino, H., 149, *164*
Hikino, Y., 149, *164*
Hillis, W. E., 41, *44*, *45*
Hillman, W. S., 194, *204*
Hinreiner, E., 49, *73*
Hoar, C. S., 127, *137*
Hochmannovà, J., 150, *164*
Hoffman, C. H., 188, *191*
Hoffman, H., 136, *137*
Hoffmann, O., 139, 144, *164*
Holzer, K., 147, *164*
Horák, M., 142, 147, *163*, *165*
Hörhammer, L., 17, 35, *45*
Horibe, I., 142, *164*
Horn, D. H. S., 41, *44*, *45*
Horowitz, R. M., 41, *45*
Horowitz, N. H., 108, *120*
Horowitz, R. M., 27, 28, *41*, *45*
Howling, D., 98, 100, *105*
Hu, S., 150, *164*
Huber, H., 125, 127, 130, 132, 135, *137*
Hulanicka, P., 95, *105*
Hutchinson, G. E., 123, *137*
Hutchinson, J., 125, *137*

I

Ichihara, K., 184, *191*
Ikeda, S., 184, *190*
Imagawa, H., 127, *138*
Ingraham, L., L., 66, *73*
Ishii, H., 150, *164*
Ivanov, D., 142, *165*

J

Jaffe, M. J., 200, 201, *204*
Jahnke, J. K., 184, 189, *190*

James, A. T., 93, 95, 98, 99, 100, 101, 102, 104, *105*, *106*
Jeffs, P. W., 116, *120*
Jizbà, J., 150, *164*
Jones, E. R. H., 110, *120*, 122, *136*
Jones, J. D., 183, *190*
Jorgensen, E. L., 17, *44*
Joseph-Nathan, P., 148, *165*
Joshi, B. S., 141, *163*
Juel, H. O., 133, *137*
Jurd, L., 2, 13, 17, 20, 27, 28, *45*

K

Kagan, H. B., 141, 144, *164*
Kagan, J., 31, 32, 35, 41, *45*
Kamada, H., 189, *191*
Kamat, V., 141, *163*
Kandler, O., 122, *137*
Kapil, R. N., 134, *137*
Kaplan, M. M., 118, *120*
Kariyone, T., 109, *120*
Kates, M., 93, *105*
Kato, E., 189, *191*
Kato, T., 144, 151, 161, *163*
Kelkar, G. R., 152, 155, 156, *164*, *165*
Kennedy, L. D., 49, *73*
Kimura, T., 127, *138*
Kinzel, H., 122, *137*
Kirchner, P. K., 152, *163*
Kirby, G. W., 113, *119*
Kitahara, Y., 144, 151, 161, *163*
Kjaer, A., 63, *73*, 110, *119*, *120*
Kleinschmidt, T., 49, *73*
Klenk, E., 99, *105*
Kluyver, A. J., 183, *190*
Knipprath, W., 99, *105*
Kobayashi, M., 144, 151, *164*
Kock, W. T., de, 144, 149, *163*
Koninszy, F. R., 188, *191*
Korte, F., 152, *164*
Kosuge, T., 66, *73*
Koukol, J., 51, 52, 55, 56, 66, *73*, 182, *190*, 203, 204
Koukol, J. E., 203, *204*
Kredich, N. M., 69, *73*
Křepinsky, J., 149, *164*
Kubitzki, K., 130, *137*
Kubota, T., 180, *190*
Kudo, S., 189, *191*
Kulkarni, G. H., 155, 156, *164*
Kumar, S., 57, *73*

Kupchan, S. M., 141, *164*
Kushnir, L. E., 155, *164*
Kuzovkov, A. D., 155, *164*

L

Laing, M., 125, *137*
Lakshmikantham, M. W., 142, 144, 156, *163*, *164*
Larabee, A. R., 92, *106*
Lawton, J. R., 175, *177*
Lebeau, J. B., 69, *74*
Ledingham, G. A., 179, 180, *191*
Leeming, P. R., 180, 187, *190*
Leete, E., 116, *120*
Lehn, J. M., 150, *165*
Leins, P., 134, *137*
Leistner, E. L., 117, *120*
Lennarz, W. J., 91, *105*
Leonhardt, R., 154, *164*
Le Quan, M., 122, *136*
Lewis, D. G., 187, *190*
Lewis, I. R., 118, *120*
Libby, P., 52, *73*
Liebman, A. A., 116, *120*
Light, R. J., 91, *105*
Ling, N. C., 111, *119*
List, P. H., 180, *190*
Locksley, J., 118, *120*
Long, L., Jr., 48, *73*
Lonroth, I., 184, *190*
Lord, K. E., 87, *90*
Lovyagina, E. V., 188, *190*
Lowe, J., 176, *177*
Lundin, R. E., 31, 32, *45*
Lynen, F., 67, *74*, 75, 89

M

Mabry, T. J., 1, 2, 3, 17, 22, 31, 32, 35, 41, 43, *45*, 111, *120*, 124, *137*, 141, 144, 162, *164*
MacDaniels, L. H., 125, *136*
McMahon, V., 100, *105*
McManus, T. T., 93, *106*
McPhail, A. T., 141, *164*
Magalhaes, M. T., 112, *120*
Mahadevan, S., 57, *73*
Majerus, P. W., 92, *105*, *106*
Mao, C-H., 49, *73*
Markham, K. R., 1, 2, 3, 17, 22, 32, 43, *45*

Marnyama, M., 108, *120*
Marshall, D., 180, 188, *190*
Marthe, J-P., 31, 41, *45*
Marumo, H., 189, *191*
Massias, M., 53, 57, *73*
Massicot, J., 31, 41, *45*
Maw, G., 176, *177*
Mayo, P., de, 142, *163*
Mazelis, M., 66, *73*
Meister, A., 183, *190*
Mentzer, C., 53, 57, *73*, 109, *120*
Merac, M., du, 154, *163*
Merkel, D., 146, *164*
Merquez, A. D., 150, 158, 162, *164*
Merxmüller, H., 134, 136, *137*
Metcalfe, C. R., 125, 127, *137*
Meyer, F., 104, *106*
Michaels, R., 70, *73*
Miljanich, P., 49, 51, 52, 55, 56, 66, *72*, 73
Miller, H. E., 141, 144, 162, *164*
Miller, S. L., 108, *120*
Minamikawa, T., 182, *190*
Minato, H., 142, 150, *164*
Miranda, E. C., 111, *119*
Mirrington, R. N., 147, *163*
Mitra, R. B., 147, *163*
Mitsuhashi, T., 184, *191*
Miura, I., 108, *120*
Mohana, Rao, P. R., 134, *137*
Molisch, H., 122, *137*
Moore, J., 118, *120*
Moore, K., 181, 182, 183, 184, 186, *191*
Moorhead, P. S., 118, *120*
Morris, L. J., 98, 100, 101, 102, 104, *105*, *106*
Mors, W. B., 112, *120*
Moscatelli, E. A., 188, *191*
Motl, O., 149, *164*
Mudd, J. B., 93, 95, *106*
Mudd, S. H., 69, *73*
Mumford, F. E., 196, *204*
Mundy, B. P., 116, *120*

N

Nagai, J., 95, 96, 101, *106*
Nakadaira, Y., 108, *120*
Nakamura, S., 144, 151, *164*
Nakanishi, K., 108, *120*
Namura, 189, *190*
Narasimhachari, N., 27, *45*

Narayanan, C. R., 147, *163*
Naves, Y. R., 155, *164*
Naya, K., 144, 151, 161, *164*, 180, *190*
Neish, A. C., 51, 55, 57, *72*, *73*, 181, 182, 183, 186, *190*, *191*
Neitz, S., 31, 41, *45*
Nichols, B. W., 99, 101, *105*, *106*
Nigam, S. N., 67, *74*
Nishimura, K., 114, *120*
Nord, F. F., 179, 180, *191*
Nosaka, S., 142, *164*
Novotný, L., 140, 149, 150, 151, 158, 160, 161, 162, *163*, *164*

O

Odham, G., 112, *120*
Ogata, K., 182, *190*
Ognjanov, I., 142, *165*
Ollis, W. D., 22, *44*
Onoda, R., 144, 151, 161, *163*
Ortiz de Montellano, P. R., 87, *89*
Osske, G., 89, *90*
Ourisson, G., 89, *90*, 150, *165*
Overath, P., 92, *106*

P

Pacheco, H., 31, 32, *45*
Pachler, K., 149, *165*
Pachler, K. G. R., 144, 149, *163*
Page, A. C., 188, *191*
Page, C. B., 122, *136*
Panizzi, L., 141, *165*
Paris, R., 180, *190*
Parker, M. W., 193, *204*
Patil, F., 150, *165*
Paul, A., 142, 156, *164*
Pfeil, E., 50, *72*
Philipson, W. R., 127, 130, *137*
Platonova, E. G., 188, *190*
Pliva, J., 142, *165*
Plouvier, V., 48, *73*
Ponsinet, G., 89, *90*
Popják, G., 75, 77, 78, 82, 87, *89*, *90*
Porter, J. W., 75, 78, 84, *90*
Power, D. M., 181, 182, 183, 186, *191*
Pridham, J. B., 62, *74*
Pritzel, E., 124, *137*
Pucher, G. W., 68, *74*
Pulle, A. A., 124, *137*

R

Rabindran, K., 147, *163*
Rajapa, S., 142, 147, 156, *163, 164*
Ramuz, H., 113, *119*
Rao, A. S., 142, 152, 155, 156, *164, 165*
Rapoport, H., 116, *120*
Rašper, V., 173, *177*
Raulais, B., 142, 156, *164*
Raulais, D., 156, *164*
Rees, H. H., 87, 89, *90*
Reichstein, T., 111, *119*
Reisch, J., 122, *137*
Renold, W., 141, 144, *164*
Ressler, C., 67, *74*
Rheinbay, J., 122, *137*
Rimington, C., 71, *74*
Rios, T., 148, *165*
Rivett, D. E. A., 150, *165*
Robbins, W. J., 188, *191*
Robinson, M. E., 47, *74*
Rode, K. M., 126, *136*
Rodionov, M. A., 93, *106*
Rodriguez, R. L., 127, *137*
Rodriguez-Hahn, L., 150, 160, *165*
Rogoff, M. H., 183, *191*
Roller, P., 111, *119*
Romaňuk, M., 155, *165*
Romo, J., 144, 148, 150, 160, *165*
Rosenthaler, L., 49, *74*, 122, 123, *137*
Rösler, H., 31, 32, 35, 41, *45*
Rosprim, L., 35, *45*
Roy, S. K., 147, *163*
Russell, D. W., 198, *204*
Russey, W. E., 87, *89*
Ryan, A. J., 184, *190*
Ryback, G., *89*

S

Sadgopal, B. S., 142, *165*
Saiki, Y., 180, 187, *191*
Sainsbury, M., 22, *44*
Sakai, A., 122, *137*
Sakamoto, Y., 183, *190, 191*
Saleh, M. R. T., 146, *165*
Saltmarsh, M. J., 62, *74*
Samek, Z., 150, 151, 152, 153, 154, 160, 161, *163, 164, 165*
Sandermann, W., 117, *120*
Santhanam, P. S., 147, *163*
Sargant, E., 177, *177*
Satoda, I., 146, *165*

Scheinmann, F., 118, *120*
Scheuerbrandt, G., 94, *106*
Schlossmann, K., 67, *74*
Schmid, J. J., 142, 147, 156, *163, 164*
Schneider, M. J., *204*
Schoes, H. K., 141, *164*
Schrader, J. C. C., 47, *74*
Schreiber, K., 89, *90*
Schroepfer, G. T., 98, *106*
Schubert, W. J., 180, 182, *191*
Schulte, E. E., 122, *137*
Schulte, K. E., 122, *137*
Schürhoff, N. N., 125, *138*
Schwarz, J. S. P., 22, *44*
Scora, R. W., 122, *137*
Seely, M. K., 49, *74*
Seikel, M. K., 35, *45*
Seshadri, T. R., 27, *45*
Shimazono, H., 180, 182, 183, *191*
Shimizu, G., 189, *191*
Shimizu, Y., *89*
Shirata, K., 144, 151, 161, *163*
Shivrina, A. N., 188, *190*
Silva, M., 131, *138*
Sim, G. A., 141, *164*
Simatupang, M. H., 117, *120*
Simes, J. J. H., 41, *44*
Smalley, H. M., 104, *105*
Smirnov, B. P., 93, *106*
Smith, B. S. W., 183, *190*
Smith, D. H., 196, *204*
Smith, G. N., 104, *105*
Smith, H., 197, 200, 203, *204*
Smith, H. G., 180, 187, *190*
Smity, M. C., 188, *191*
Snatzke, G., 153, *165*
Sorensen, N. A., 110, *120*
Šorm, F., 140, 141, 142, 146, 147, 149, 150, 151, 152, 153, 154, 155, 158, 160, 162, *163, 164, 165*
Souček, M., 141, 147, *163, 165*
Spencer, D. H., 122, *136*
Spencer, J. D., 116, *119*
Spenser, I. D., 53, 65, 66, *72, 74*
Spitzer, J. C., 140, 142, 146, *165*
Sporne, K. R., 176, *177*
Squires, C., 91, *105*
Stafford, H. A., 203, *204*
Stahl, E., 162, *165*
Stary, F., 150, 158, 162, *164*
Steelink, C., 140, 142, 146, *165*
Steglich, W., 113, *119*
Stern, D. J., 13, 32, *45*

Stevens, D. L., 70, *74*
Stevens, R. L., 66, *74*
Stoker, J. R., 71, *72*
Stoll, A., 150, *164*
Stowe, B. B., 196, *204*
Strasser, R., 17, *45*
Strobel, G. A., 69, 70, *72*, *74*
Stumpf, P. K., 91, 92, 93, 95, 100, 102, *105*, *106*
Subba Rao, P. V., 181, 182, 183, 184, 186, *190*, *191*
Suchý, M., 146, 152, 153, 154, 155, 156, *165*
Sudarsanam, V., 142, 156, *164*
Sumere-de-Preter, C., van, 179, 180, *191*
Sumi, M., 154, *165*
Sumi, Y., 156, *164*
Swain, T., 20, *45*, 109, *120*
Swift, III, W. T., 22, *45*

T

Tagaki, I., 144, 151, 161, *164*
Takara, K., 127, *138*
Takeoshi, T., 150, *165*
Takhtajan, A., 124, *138*
Talamo, B., 92, *105*
Tamahashi, Y., 150, *165*
Tamelen, E. E., van, 87, *90*
Tanada, T., 203, *204*
Tanenbaum, S. W., 184, *190*
Tapper, B. A., 57, 58, 59, 60, 63, 64, 65, *74*, *73*
Taves, C., 122, *137*
Tchii, S., *89*
Terahara, A., 108, *120*
Tether, L. R., 144, 158, *163*, *164*
Thaller, V., 122, *136*
Thomas, G. M., 113, *119*
Thomas, M. B., 1, 2, 3, 32, 43, *45*
Thomas, R. G., 197, *204*
Thomson, R. H., 117, *119*
Thorn, G. D., 70, *74*
Tieghem, Ph., van, 125, *138*
Tirimanna, A. S. L., 67, *72*, *73*
Tochikura, T., 182, *190*
Toman, J., 150, 151, 158, 162, *164*
Tomioka, K., 189, *191*
Tomkins, G. M., 69, *73*
Toribio, F. P., 156, 158, *163*
Towers, G. H. N., 55, 57, *72*, 108, *120*, 181, 182, 183, 184, 186, 188, *190*, *191*

Tozyo, T., 150, *164*
Trecanni, V., 183, *191*
Treharne, K. J., 84, *89*
Tschiersch, B., 53, 63, 66, 68, 71, *74*
Tse, A., 108, *120*
Turley, R. J., 158, *163*
Turner, B. L., 109, *119*, 122, 123, 124, *136*, *137*, 162, *164*

U

Uchiyama, K., 182, *190*
Ueno, A., 180, 187, *191*
Underhill, E. W., 63, 64, 65, 66, *74*
Uribe, E., 52, 55, 56, *74*
Uritani, I., 182, *190*

V

Vagelos, P. R., 91, 92, *105*, *106*
Van Sumere, C. F., 179, 180, *191*
Vickery, H. B., 68, *74*
Villiers, J. P., 149, *165*
Vining, L. C., 179, 180, *191*
Viswanathan, N., 142, 147, *163*, *164*
Vitucci, J. C. 179, 180, *190*
Vivar, A. R., de, 144, 148, *165*

W

Waaler, T., 149, 150, 162, *163*
Wagner, H., 35, *45*
Wain, R. L., 122, *136*
Waiss, A. C., 31, 32, *45*
Wakaki, S., 189, *191*
Walland, A., 122, *137*
Ward, E. W. B., 69, 70, *74*
Warrington, B. H., 118, *120*
Weatherston, J., 112, *120*
Webb, L. J., 122, 123, *138*
Webb, J. P., 102, *105*
Weintraub, R. L., 197, *204*
Weiss, S. K., 111, *119*
Went, F. W., 193, *204*
Wetter, L. R., 63, 64, *74*
Wetterstein, R., 124, *138*
Whiting, M. C., 180, 188, *190*
Wieffering, J. H., 127, 129, 133, *138*
Wilcox, M. E., 111, *120*
Willett, J. D., 87, *90*
Williams, R. J. H., 78, 79, 80, 82, 84, *89* *90*

Willis, J. C., 167, 172, 176, *177*
Wilson, D. V., 187, *190*
Wilson, E. E., 66, *73*
Wolfrom, M. L., 118, *120*
Woods, M. C., 108, *120*, 144, 151, 161, *163*
Woolard, G. R., 150, *165*
Wyler, H., 111, *120*

Y

Yamada, H., 182, *190*
Yamada, M., 102, *106*

Yoshida, N., 146, *165*
Yoshii, E., 164, *165*
Yosioka, I., 127, *138*
Yosioka, T., 149, *164*
Young, M. R., 181, *191*
Yuan, C., 100, *106*

Z

Zalkow, L. H., 150, *164*
Zenk, M. H., 117, *120*
Zijp, J. C. M., van, 183, *190*
Zinke, A., 147, *164*

Chemical Compounds Index

(Structural formulae for compounds are given on those pages indicated by a number in bold type)

A

Acacetin, 5
Acacipetalin, 48, 71
Acetylene compounds, 102, 104
Actinomycin D, 201
Adenostylone, 150, **160**
Adonitol, 122
Afrormosin, 22, 24
Aglycones, *see also specific names*, 47, 48
Alanine, 69–70
Albicolide, **152**, 154
Albopetasin, 150, 151
Albopetasol, 150
Aldoximes, *see also specific names*, 58, 62, 64–65
Alizarin, **117**
Alkaloids, *see also specific names*, 111, 112, 115, 116, 122, 123, 130, 139, 158
Aluminium, 122
Ambrosanolides, *see also* Pseudo-guaianolides, 142, **143**, 145, 147, 148
Ambrosin, 157, 158
Ambrosiol, 162
Amentoflavone, 7
Amino acids, *see under individual names*,
γ-Aminobutyric acid, 70
4-Amino-4-cyanobutyric acid, 69
2-Amino-4-phenylbutyric acid, 64
α-Aminopropionitrile, 69
Amygdalin, 48
Anabasine, 116
Androcymbine, **115**
Angelyljaponicin, 150, 151, **159**
Anthocyanidins, 170, 174
Anthocyanins, 17
Anthranilic acid, 125, 131
Anthraquinones, 117, 123
Apigenin, 5, **12**, 13, **19**
Apigenin, 7-*O*-glucoside, 5, **16**
  —7-*O*-neohesperidoside, **42**
Apigenin 4′-methyl ether, 5
  8*

Arctiopicrin, **152**
Aristolochia, lactone, **141**
Artemisin, 154, **155**
Asparagine, 66, 67–68, 71–72
Asperuloside, 131
Athrotaxin, 114, **115**
Aucubin, 127, 128, 129, **130**
Aurones, 2
Auxins, 194, 198, 200

B

Baicalein, 5, 14
Bakkenolides, **143**, 144
Bakkenolide, A, 150, **161**
Bakkenolides, B, C, D, 151
Balchanin, **155**, 156
Barrigenol, 131
Bayin, 4, **19**
Benzaldehyde, 49, 50
Benzaldehyde cyanohydrin, 49, 50
Benzoic acid, 186, 187, 189
Benzyl glucosinolate, 63, **64**, 65
Benzylisoquinoline, 113
Bergaptene, 125, 131
Betacyanins, 124
Betalains, 111, 112, 124
Betanin, **111**
Betaxanthins, 124
Bile acids, 108
Bile salts, 112
Biochanin A, 23
Bisabolanes, 144
Branched chain fatty acids, *see also under specific names*, 93, 112

C

Caffeic acid, 122, 125, 127, 128, 129, 139, **171**, 172, 173, 174, 175, 180, 182, **183**, 186, 188
Camptothecin, **130**

Carabrone, 142, **143**
Cardenolide glycosides, 110
Cardinanes, 144
Carotene
  α-Carotene, 78, 80, **81**, 82, **83**, 84
  β-Carotene, 78, 79, 80, **81**, 82, **83**, 84
  γ-Carotene, 79, 80, 83, 84
  δ-Carotene, 79, 80, 83, 84
  ε-Carotene, 80
  ζ-Carotene, 80, **83**
Carotenes, cyclic, 78–82
Carotenoids, *see also specific names*
  79–90, 193
Catalpol, 133
Centaurein, 11
Chalcones, 2, 28 171
Chamazulene, 140
Chlorogenic acid, 125, 127, 180
Cholesterol, 116, 117
Chromoalkaloids, 124
Chrysoeriol, 6
Cineol, 122
Cinnamaldehyde, 180
Cinnamic acid, cinnamate, 171, 172, 179,
  180, 181, **182**, **183**, 184, **186**, 187, 188,
  189, 203
Cnicin, **152**
Codeine, 113, **114**
Codeinone, 113, **114**
Colchicine, 114, **115**
Coniferyl alchohol, 181
Coniine, 116, **117**
Cornin, 131
Coronopilin, 157, 158, 162
Cortisalin, 180, 188, 189
Costunolactone, **155**
Costunolide, **152**, 154, 156
p-Coumaric acid, **171**, 180, 181, 182, **183**,
  186, 187, 188
Coumarins, 122, 180
p-Coumaryl alcohol, 181
Crepenynic acid, **103**, 104
Cyanidin, 174
β-Cyanoalanine, 67, 68, 70, 71
Cyanobutyric acid 70
Cyanogenic glucosides, *see also specific*
  names, 47–74, 50–57, 57–63, 70
Cyanohydrins, *see also specific names*,
  47, 50, 58, 59, 60, 61, 62
Cyclic carotenes, 78–82
Cycloartenol, 85, 86, 87, 88, 89
Cynaropicrin, **155**
Cysteine, 68, 69
Cytokinins, 200

D

Daidzein, 22, 23
Damsin, 157
Decanal, 125
Decanoic acid, decanoate, 93, 94, 97
Decompostin, 150, **160**
Dehydrocostus lactone, **155**
Delphinidin, 174
4-Demethyl phytosterol, 89
Deoxylapachol, **117**
Dhurrin, 48, **49**, 50, 51, 52, **54**, 55, 56, 66,
  70
Diangelyljaponicin, 151, **159**, 160
Dienoic acids, 101
Dihydrofisetin, 25
Dihydroflavonols, 2, **22**, 20–31, **40**
Dihydrophytoene, 84
Dihydroquercetin, **21**, **26**, **30**
Dihydrorobinetin, 25
3,4-Dihydroxybenzoic acid, 184
2,5-Dihydroxybenzoylformic acid, 184,
  185
3′,4′-Dihydroxyflavone, 4
4′,7-Dihydroxyflavone, 4
4′,7-Dihydroxyisoflavone, 23
5,7-Dihydroxyflavone, 14
5,7-Dihydroxyisoflavone, 22
4′,5-Dihydroxy-7-methoxyflavanone
  5-O-glucoside, 25
Dihydroxymandelic acids, 184, **185**
Dihydroxyphenylacetic acids, 184
3,4-Dihydroxyphenylpyruvic acid, 184
3′,5′-Dimethoxy-4′,5,7-trihydroxyfla-
  vone, 6
Dimethylallyl pyrophosphate, 75, **76**
β-Dimethyl-α-hydroxyacrylonitrile-β-D-
  glycoside, 71
Dinatin, 133
Diosmetin triacetate, 14, **15**
Diosmin, 42
DMAPP, *see* Dimethylallyl pyrophosphate

E

Elephantin, **141**
Ellagic acid, 126, 127, 128, 129, 131, **171**,
  172, 176
Ellagitannins, 127, 131, 176
Emetine, **130**
Emodin, 117
Endocrocin, 117
Eremophilanes, 140, **149**, 150, 158–160

Eremophilanolides, 142, **143**, 145, 149
Eremophilene, **149**, 150
Eremophilenolide, 150
Eriodictyol, **21**, 25, **26**, **30**
Esculetin, 122
Eudesmanolides, *see also* Santanolides, 142, **143**
Eugenol, 122
Euryopsol, 150
Euryopsonol, 150

F

Farnesyl pyrophosphate, 76, 77, 142, 143
Fatty acids, *see also specific names*, 91–105
Ferulic acid, 125, **171**, 172, 174, 175, 188
Fisetin, 8
  -3-*O*-glucoside, 9, 17
Flavandiols, 169, **170**, 172, 175
Flavanones, 20–31, 171
Flavonoid glycosides, 31
Flavonoids, 1–45, 111, 167–177, 181, 188, 189, 193–204
Flavone glycosides, 112, 133
  Flavones, **2**–20, 171, 173
Flavonols, **2**, 20–21, 125, 139, 169, **170**, 171, 172, 173, 174, 175
FPP, *see* Farnesylpyrophosphate
Formononetin, 23
Fructose, 199
Fukinolides, **143**, 144, 151
Furanoeremophilanes, 162
Furanoligularenone, 150
Furanopetasin, 151, 158, **159**, 160
Furoeremophilane, 150, 151
Furoeremophilone, 151

G

Gafrinin, **143**
Galangin, 14, 31, 32
Gallic acid, 126, 127, 128, 129, 131
Gallitannins, 127, 131
Geigerin, 140
Geigerinin, **148**
Genistein, **21**, 22, 23
  -7-*O*-glucoside, 23
  -4′-methyl ether, 23
  -7-methyl ether, 23
  -7-*O*-rhamnoglucoside, **28**
Gentisic acid, 183, 184, **185**
Geranyl pyrophosphate, 76, 76–77
Geranylgeranyl pyroposphate, **83**

Germacranolides, 142, **143**, 145, 146, 148, 152, 156
Gibberellic acid, gibberellin, 200
Ginkgolides, 108
Gluconasturtiin, 64
Glucoputranjivin, 64
Glucose, 180, 182, 199
Glucosides, *see under specific names, also* Cyanogenic-, Iridoid, etc.
Glucosinolates, 63–66, **110**
6-*C*-Glucosylapigenin, 5
8-*C*-Glucosyl-4′,7-dihydroxyflavone, 4
p-Glucosyloxybenzaldehyde, 52
8-*C*-Glucosyl-3′,4′,5,7-tetrahydroxyflavone, 6
Glucotropaeolin, 63, 64
Glutamic acid, 69
Glutamine, 71–72
γ-Glutamyl-β-cyanoalanine, 67, 68
Glycoflavones, 171, 173
Glycozolidine, 117
Glycozoline, 117
Gossypetin, 174
Guaiacyl derivatives, 188, 189
Guaianolides, 142, **143**, 145, 146, 147, 148, 154
Gynocardin, 48

H

Haemanthamine, 115, **116**
Hedacidin, 66
Helenalin, 140, 146, 157, 158
Heptaphylline, 117
Hesperidin, 25, **33**
7-*cis*-Hexadecenoic acid, 99
9-*cis*-Hexadecenoic acid, 99
3-*trans*-Hexadecenoic acid, 99–100, 102
2-(3-*trans*) Hexadecenoyl-phosphatidyl glycerol, 100
Hexaglucoside derivatives, 196
Hexahydroxydiphenic acid, **176**
Hexitols, 122, 133
Hispidin, 180, 187, 188, 189
Homogentisic acid, 183, **184**, **185**
Homoplantaginin, 133
p-Hydroxybenzaldehyde, 52
4-Hydroxybenzoic acid, 184, **185**, **186**, 187
4-Hydroxybenzoylformic acid, 184, **185**
Hydroxycinnamic acids, 122
4-Hydroxycoumarin, 186
6-Hydroxyeremophilenolide, 150, 151

5-Hydroxyflavone, 14
7-Hydroxyflavone, **3**, 14
3-Hydroxyflavones, 14
9-Hydroxyfuroeremophilane, 150
9-Hydroxyfuroeremophilene, 158, **159**
6-Hydroxygenistein, 22, 27
N-Hydroxyglycine, 66
α-Hydroxyisobutyric acid, 53
α-Hydroxybutyronitrile, 59, 60
α-Hydroxyisobutyronitrile-β-gluco-
  pyranoside, 47, 48
4-Hydroxymandelic acid, 184, **185**
p-Hydroxy-L-mandelonitrile, 48, 49
  -β-glucopyranoside, 49
m-Hydroxymandelonitrile-β-D-gluco-
  side, 71
o-Hydroxyphenylacetic acid, 184
p-Hydroxyphenylacetic acid, 183, 184,
  185, 186
α-Hydroxyphenylacetonitrile, 59, 60
m-Hydroxyphenylalanine, 183
N-Hydroxyphenylalanine, 66
p-Hydroxyphenyl-DL-lactic acid, 52
p-Hydroxyphenylpyruvic acid, 52, 183,
  184, 185
D-12-Hydroxy-9-octadecenoic acid, 102
Hydroxysugiresinol, 114, **115**
DL-β-Hydroxyvaline, 53
Hymenoxin, 7

I

Ilicic acid, **158**
Indicamin, 153
3-Indole acetaldoxime, 57
3-Indole acetonitrile, 57
3-Indole acetic acid, 194, 195, 198, 203
IPP, *see* Isopentanyl pyrophosphate,
Iresin, **141**
Iridoid glucosides, 126, 127, 128, 129,
  131, 133
Irigenin, 24
Isoadenostylone, 150, **160**
Isobutyraldoxime, 57, 58, **59**, 63, 64–65
  -glycoside, 63
Isobutyronitrile, 59, 60
Isocitric acid, 122
Isoferulic acid, 182, **183**
Isoflavones, 2, **22**
Isoleucine, 53, 54, 70, 71
Isoorientin, 6
Isopentenyl pyrophosphate, 75, 76
Isopetasin, 150
Isopropyl glucosinolate, 63, **64**, 65

Isorhamnetin, 10
  -3-O-rutinoside, 10
Isovitexin, 5
Ivalbin, 142, **143**
Ivangulin, **143**

J

Jacein, 11
Junipal, 110
Jurineolide, **153**

K

Kablicin, 151, **159**, 160
Kablikopetasin, 151
Kaempferol, 7, 14, 125, 169, **170**, 175, 198,
  200
  -4'-methyl ether, 8
  -7-O-neohesperidoside, 8
  -3-O-robinoside 7-O-rhamnosylgluco-
    side, 8
  -3-triglucoside, **196**, 197, 198, 200
Keto acids, *see under specific names*,
9-Ketofuroeremophilane, 151
9-Ketofuroeremophilene, 158, **159**
α-Ketoisovaleric acid oxime, 57–58
α-Ketoximes, 54, 65, 66

L

Lactucin, 140, **147**
Lactucopicrin, 140, **147**
Lanosterol, 85, **86**, 87, **88**, 89
Leucoanthocyanins, 125, 127, 170, 172,
  173, 174, 175, 177
Leucocyanidin, 125, **170**, 172, 173, 174,
  175
Leucodelphinidin, **170**, 174, 175
Lignans, 109
Lignins, 109, 179, 181, 187–188, 189, 203
Ligularol, 150
Ligularon, 150
Linamarin, 47, 48, **49**, 53, **54**, 55–64, 70,
  71
Linoleic acid, 96, 100, 102, **103**
Linolenic acid, **103**
1-Linolenoyl-2-(3-*trans*-hexadecenoyl)-
  phosphatidyl glycerol, **99**
Loganic acid, **130**
Loganin, 127, 129, **130**, 131
Lotaustralin, 47, 48, 53, **54**, 70, 71
Lotusin, 71

Lumisantonin, 142, **143**, 146
Luteolin, 6, **12**, 13, **19**, 35, **36**, 172, 173
Luteolinidin, **173**
Lycopene, 78, 79, 80, 82, **83**, 84
Lycopersene, 84
Lysine, 116

M

Maltose, 199
Mannitol, 122
Mesembrine, 115, **116**
Methoxycinnamic acid, 174, **183**
Methoxydihydrocostunolide, **155**, 156
4′-Methoxy-6,7-dihydroxyisoflavone, 24
4′-Methoxyflavone, 15
4′-Methoxy-7-hydroxyisoflavone, 23
3′-Methoxy-4′,5,7-trihydroxyflavone, 6
Methylcatalpol, 133
Methyl cinnamate, 186
Methyl *p*-coumarate, 180, 182
Methyl *p*-methoxycinnamate, 180, 182
24-Methylenecycloartenol, 87
N-Methylisopelletierine, **116**
Mevalonic acid, 75–90
Mexicanin E, **143**, 144
Mikanolide, **147**
Monoenes, *see also specific names*, 93–98, 101
Monotropeoside, 131
Morphine, 113, **114**
Murrayanine, 117, **118**
Mycomycin, 110
Myricetin, 14, **170**, 175
Myristic acid, 101

N

Narigenin, 25
Neoadenostylone, 150, **160**
Neurosporene, 78, **83**
Nevadensin, 7
Nicotine, 116, 122
Nitriles, *see also specific names*, 53, 54, 60, 65, 69
Nitrobenzene, 188
Norwogonin, 5, 14

O

Cis-9-Octadecen-12-ynoic acid, 103, 104
trans-11-Octadecen-9-ynoic acid, 104
Octanoate, octanoic acid, 93, 94, 97
Oleanolic acid, 131, 133
Oleic acid, **94**, 96, 100, 101, 102, **103**, 104

Onopordopicrin, **153**
Orobol, 24, **26**, **30**
   -7-*O*-glucoside, 24, **28**
   -4′-methyl ether, 24
Oximes, 57, 62–63
2-Oximinovaline, 57

P

Palmitate, palmitic acid, 96, 99, 100, 101
Paracotoin, 108
Parkeol, 87, **88**, 89
Parthenin, 157, 158, 162
Parthenolide, **141**
Patulitrin, 11
Penduletin, 11
3,3′,4′,5′,7-Pentahydroxyflavanone, 25
3,3′,4,5′,7-Pentahydroxyflavone, 10
3,3′,4′,5,7-Pentahydroxyflavone, 9
3,3′,4′,5,7-Pentahydroxy-6-methoxyflavone 7-*O*-glucoside, 11
Pentitols, 122
Petasalbin, 160, 161
Petasin, 150
Petasitin, 151, **161**
Petasolides A, B, 150
Phenolic acids, *see also specific names*, 127, 128–129, 203
Phenols, 111, 197
Phenylacetaldoxime, 58, 64, 65, 66
Phenylacetic acid, 184, **186**
Phenylacetonitrile, 59, 60
Phenylalanine, 53, 54, 58, 59, 60, 63–64, 65, 71, 180, 181, 182, **183**, 184, **186**, 187, 188, 189, **203**
γ-Phenylbutyrine, 64
6-Phenylcoumalin, 108
Phenylcrotonaldehyde, 180
Phenyllactic acid, 184, **186**
Phenylpropane, 180, **183**
Phenylpyruvic acid, 186
   -oxime, 58
Pheromones, 112
Phloretic acids, 182, **183**
Phytochrome, 193, 195, 197, 198, 199, 201, 203
Phytoene, 77–85, **83**
Phytofluene, **83**, 84
Phytomelanes, 139
Phytosterols, 85–89, 116
Pinene, 122
Pinnatifidine, **147**
Pinocembrin, 25
Pinoresinol, 109

Plantaginin, 133
Plantagonin, 133
Planteose, 133
Polyacetylenes, 109–110, 122, 126
Polyenoic acids, *see also specific names*,
  103, 104
Pratensein, 24
Protocatechuic acid, **185**, **186**, 187, 189
Protochlorophyll, 193
Prulaurasin, 48
Prunasin, 48, 53, **54**, 58, 59, 60, 62, 70, 71
Prunetin, 23
Pseudoguaianolides, *see also* Ambros-
  anolides, 142, **143**, 147, 156, 158, 162
Psilostachyin, A, **141**, 144, 162
Psilostachyin, B, C, **143**, 144, 162
Psilotin, **108**

Q

Quercetagetin, **174**,
Quercetin, 9, **12**, **16**, 14, 19, 125
  -3-(*p*-coumaroyltriglucoside), **196**,197,
    200, 201, 202
Quercetin conjugates, 169, **170**, 175, **201**
  -3,7-*O*-diglucoside, 9
  -3′-methyl ether, 10
  -7-methyl ether, 9
  -4′-methyl ether 7-*O*-rutinoside, 10
  -3-*O*-rhamnoside, 9, 12, 16
  -3-triglucoside, **196**, 197, 200
Quercetrin, 9, 12, 16
Quinic acid, 122, 126, 128

R

(−)-Reticuline, 113, **114**
Rhamnetin, 9
Ribose, 199
Ricinoleic acid, 102, **103**
Robinetin, 10
Robinin, 8

S

Sakuranin, 25, **29**
Salonitenolide, **153**
Salonitolide, **153**
(+)-Salutaridine, 113, **114**
Sambunigrin, 48
Santanolides, *see also* Eudesmanolides,
  142, **143**, 145, 146, 147, 148, 156
Santonin, 140, 146
Saponins, 125, 131
Saurin, **155**
Saussurea lactone, 142, **143**, **155**, 156

Scabiolide, **153**
Scopoletin, 122
Scutellarein, 133
Sedamine, **116**
Senecio alkaloids, 158
Serine, 68
Sesquiterpene lactones, 142–144, 144–
  158
Sesquiterpenoids, *see also specific names*,
  139–165
Shikimic acid, 51, 117, 118, 122, 184
Silicic acid, 122
Sinapic acid, **171**, 172, 174
Sinapyl alcohol, 181
Sorbitol, 122, 133
Spathulin, 157, 158
Squalene, 77, 78, 79, 80, 82, 84, 85, **86**, 87,
  88
Stachyose, 122, 133
Stearic acid, 96 101, **103**
Sterculic acid, 99, **100**
Stevin, **148**
Styrylpyrone, 188
Sucrose, 199, 200, 201
Syringaldehyde, 188
Syringyl derivatives, 188, 189

T

Tamarixetin 7-*O*-rutinoside, 10
Taxifolin, 25
Taxiphyllin, 48, 55, 57, 70
Tectoquinone, **117**, 118
Temisin, 142, **143**
Tenulin, 146, **157**, 158
Terpenaids, *see under group and specific
  names*
Terthienyl, 110
Tetradecanoate, 93
3,3′,4,7-Tetrahydroxyflavanone, 25
3′,4′,5,7-Tetrahydroxyflavanone, 25
3,3′,4′,7-Tetrahydroxyflavone, 8
3,4′,5,7-Tetrahydroxyflavone, 7
3′,4′,5,7-Tetrahydroxyflavone, 6
3′,4′,5,7-Tetrahydroxyisoflavone, 22, 24
4′,5,6,7-Tetrahydroxyisoflavone, 27
3′, 4′, 6, 8-Tetramethoxy-5, 7-dihydroxy-
  flavone, 7
Texasin, 22, 24, **26**, **30**
  -7-*O*-glucoside, 24
  -6-methyl ether, 24
Thebaine, 113, **114**
Thymol, 122

Tricin, 6
3,4,5-Trihydroxycinnamic acid, **171**
4',5,7-Trihydroxyflavanone, 25
5,6,7-Trihydroxyflavanone, 29
3',4',7-Trihydroxyflavone, 4
3,4',7-Trihydroxyflavone, 7
3,5,7-Trihydroxyflavone, 31
4',5,7-Trihydroxyflavone, 5
5,6,7-Trihydroxyflavone, 5
5,7,8-Trihydroxyflavone, 5
4',5,7-Trihydroxyisoflavone, 22, 23
3',5,7-Trihydroxy-4'-methoxyflavanone
7-*O*-rutinoside, 25
3',6,7-Trimethoxy-4',5-dihydroxyfla-
vone, 11
4',6,8-Trimethoxy-5,7-dihydroxyflavone,
7
3, 3', 6-Trimethoxy-4', 5, 7-trihydroxy-
vone 7-*O*-glucoside, 11
3, 4', 6-Trimethoxy-3', 5, 7-trihydroxyfla-
vone 7-*O*-glucoside, 11
3', 4',6-Trimethoxy-5,5',7-trihydroxyiso-
flavone, 24
Triterpene carboxylic acids, 133
Triterpenes, 75–90, 111, 122, 125, 131
Tropolones, 111, 114
Tubiferin, **147**
Tyrosine, 51, 54, 55–57, 113, 180, 181,
182, 183, **184**, **185** 187, 198

U

Umbelliferone, 180
Ursolic acid, 133

V

cis-Vaccenic acid, 94
Valine, 53, 54–55, 60–65, 70, 71
Vanillin, 188
Verbascose, 122
Verbenalic acid, 129, **130**
Verbenalin, 127, **130**
Vermeerin, 144, **148**, 149
Vicianin, 48, 71
Virginolide, **147**

X

Xanthinin, 142, **143**
Xanthones, 118
Xanthurin, 142, **143**
Ximenynic acid, 104
Xylose, 186

Z

Zaluzanin, A, B, **148**
α-Zeacarotene, 80, **83**
β-Zeacarotene, **83**
Zierin, 48, 71

Genus and Species Index

A

Abrotanum, 156
Acacia, 48
Achillea millefolium, 162
Adenostyles, 160, 161
  alliariae, 150, 160
Aesculus, 122
Agavales, 168
Aizoaceae, 115, 116
Alangiaceae, 126, 128, 132, 134
Alangium, 127, 130
  rotundifolium, 128
Albizzia, 201
Alismataceae, 177
Alismatales, 168, 169
Allioideae, 174
Aloe, 122
Alstroemeriales, 168
Amanita citrina, 180
Amaryllidaceae, 115, 116
Amaryllidales, 168
Ambrosia, 147, 157, 158
Ambrosia cumanensis, 162
  ilicifolia, 158
  psilostachyia, 144, 162
Ambrosiaceae, 144, 148, 154
Anabaena variabilis, 95
Anabasis aphylla, 116
Androcymbium melanthioides, 115
Andropogoneae, 173
Aniba, 108
Annoaceae, 113
Annona reticulata, 113
Anthemideae, 145, 146, 154, 156
Apocyanaceae, 110, 111
Aponogetonales, 168
Araceae, 173, 174
Arales, 168, 169
Araliaceae, 122, 125, 126, 127, 130, 131, 132, 135
Arctium lappa, 152
  minus, 152
  nemorosum, 152
  tomentosum, 152

Arctotidae, 145, 146
Arecales, 169
Argophyllum, 130
Aristolochiaceae, 170
Aristolochiales, 169
Artemisia, 145, 156
  balchanorum, 156
  kurramensis, 146
Asclepiadaceae, 111
Ascomycetes, 184
Aspidosperma, 111
Asteraceae, 139–165
Asterales, 132
Astereae, 145, 146
Asteroideae, 144
Athrotaxis selaginoides, 144
Aucuba, 127, 130, 131
  japonica, 129

B

Bacopa, 122
Balduina, 157, 158
Basidiomycetes, 179–161
Batis, 124
Berberidales, 168
Bignoniaceae, 117
Bixales, 125, 134
Boletus scaber, 180
Bromeliaceae, 174
Bromeliales, 168, 169
Bryophyta, 180
Burmanniaceae, 173
Burmanniales, 168
Butomaceae, 172
Butomales, 168

C

Cacalia, 160, 161
  decomposita, 150, 160
Cactaceae, 124
Calendula officinalis, 146
Calenduleae, 145, 146
Callitrichaceae, 133
Callitriche, 133, 134

Calyciferae, 168
Camptotheca, 127, 130
  acuminata, 128
Cannaceae, 175
Capparidaceae, 110, 134
Capparidales, 134
Caricoideae, 174
Carpesium abrotaniodes, 142
Carpobrotis chilense, 98
Caryophyllaceae, 124
Casuarinaceae, 170
Centaurea calcitrapa, 152
  diffusa, 152
  iberica, 152
  micranthos, 152
  ovina, 152
  salonitana, 153
  scabiosa, 153
  stoebe, 152
Centrospermae, 11, 115, 124
Cephalotus, 135
Ceratocystis fimbriata, 180
Chenopodiaceae, 116, 124
Chlorella pyrenoidosa, 67
  vulgaris, 95, 98, 99, 100, 101
Chromobacterium violaceum, 70
Chrysanthemum parthenium, 141
Cichorieae, 145, 147
Cichorioideae, 144
Cicuta virosa, 110
Cistales, 134
Citriobatus, 124
Clausena heptaphylla, 117
Claviceps purpurea, 102
Cnicus benedictus, 152
Cochlearia officinalis, 64
Colchicum, 114, 115
  autumnale, 115
Commelinaceae, 174
Commelinales, 186, 169
Compositae, 110, 112, 132, 139–165, 170
Conyza dioscoridis, 146
Conium maculatum, 116
Cornus, 127
  mas, 129
Cornaceae, 126–131, 132, 134
Cornales, 126–131, 132, 135
Corokia, 127, 130, 131
  cotoneaster, 129
  virgata, 129
Corticium salicinum, 180, 188
Corynebacterium diphtheriae, 98

Crassulaceae, 116
Crepis rubra, 104
Crocus, 175, 176
Croton salutaris, 113
Cruciferae, 110, 134
Curtisia, 130
Cynara, 146
  cardunculus, 155
  scolymus, 155
Cynareae, 145, 146, 152, 154, 154–156
Cyanastraceae, 174
Cyclanthaceae, 173
Cyclanthales, 168, 169
Cyperaceae, 173, 174
Cyperales, 168, 169

D

Davidia, 127
  involucrata, 128
Davidiaceae, 126
Dendrobiae, 175
Dicotyledoneae, 168, 169
Didieraceae, 124
Dioscorea, 175
Dioscoreaceae, 173, 175, 1776
Dioscoreales, 168, 169
Dipsacales, 132
Dipterocarpaceae, 112

E

Ebenaceae, 170
Enantioblastae, 168
Equisetophyta, 180
Ericaulaceae, 174
Eriocaulales, 168, 169
Eriocaulon, 174
Escallonia, 126, 131
Escalloniaceae, 130
Escherichia coli, 67, 68, 91, 92
Eucalyptus, 48
Euglena, 95, 101
Eupatorieae, 145, 148, 160
Euphorbia, 87
Euphorbiaceae, 113
Euryops floribundus, 150

F

Fistulina hepatica, 180
Flagellariaceae, 174
Fomes applanatus, 188
  pinicola, 188

Franseria, 147
  dumosa, 158
Fumariaceae, 134
Fusarium, 70

G

Gaillardia, 156, 157, 158
  fastigiata, 147
  grandiflora, 157
Galium, 117
Garrya, 127, 130, 131, 134, 135
  elliptica, 128
Garryaceae, 126, 128
Geigeria africana, 144, 149
  aspera, 148, 149
Gentianaceae, 128
Gentianales, 131, 132
Ginkgo biloba, 108
Glumiflorea, 168
Glycosmis pentaphylla, 117
Graminales, 168
Gramineae, 172, 173, 174, 175
Griselinia, 127, 130, 131
  littoralis, 129
Gunnera, 135
Guttiferales, 134
Gynandrae, 168
Gynocardia odorata, 48

H

Haemodorales, 168, 169
Haloragaceae, 133
Hamamelidales, 135
Hedera helix, 126
Heleniae, 145, 147, 154, 156–158
Helenium, 146, 156, 157, 158
  mexicaneum, 144
  pennatifidum, 147
  virginicum, 147
Heliantheae, 144, 145, 147, 148, 149, 154, 156–158
Helobiae, 168
Helwingia, 127, 130
  japonica, 129
Herpestis, 122
Hippuridaceae, 133–134
  Hippuris, 133, 134
  vulgaris, 133
Homogyne, 161
  alpina, 150, 161
Hydrangea, 126, 131
Hydrangeoideae, 132
Hydrocharetaceae, 172, 173

Hydrocharetales, 169
Hymenoclea, 157, 158
  salsola, 158
Hymenophyllum, 124
Hypericum, 112

I

Inuleae, 145, 148, 149
Iridaceae 173, 174, 175
Iridales, 168, 169
Iris, 175
Iva, 142, 147, 156, 157, 158

J

Juglans, 122
Juncaceae, 174
Juncaginales, 168
Juncales, 168, 169
Jurinea, 154
  albicaulis, 152
  cyanoides, 153

K

Kaliphora, 130
Kielmeyera, 118

L

Lathyrus, 71
Latimeria, 108
Lauraceae, 108
Laurales, 169
Leguminosae, 110, 112, 132
Leguminosales, 135
Lemnaceae, 174
Lentinus lepideus, 179, 180, 182, 183, 186
Lepidium sativum, 64
Ligularia fisheri, 150
Liguliflorea, 147
Liliaceae, 172, 173, 174, 177
Liliales, 168, 169
Lilliflorea, 168, 175, 176
Lilioideae, 174
Linum usitatissimum, 48, 49
Liquidambar, 131
Litorella, 133, 134
  uniflora, 133
Lotus, 48
  arabicus, 71
  tenuis, 68
Lupinus angustifolius, 68
Lycopodophyta, 180

M

Maclura, 118
Magnolia grandiflora, 141
Magnoliaceae, 141
Magnoliales, 132, 169
Marantaceae, 173, 175
Mastixia, 127, 130
  arborae, 129
  pentandra, 129
  rostrata, 129
  trichotoma, 129
Mastixiaceae, 126
Mayacaceae, 174
Melanophylla, 130
Mentha, 113
Mesembryanthemum, 115
Mesua, 118
Michelia champaca, 141
Mimosa, 201
Molluginaceae, 124
Monocotyledoneae, 168, 169, 173
Moraceae, 118
Moringaceae, 110
Mutisieae, 145, 146
Murraya koenigii, 117
Musaceae, 173, 175

N

Nagadales, 168
Nandina domestica, 71
Nasturtium officinale, 64, 66
Nectandra, 101
Nicotiana, 116
Nymphaeales, 169
Nyssa, 127, 130
  javanica, 128
  silvatica, 128
Nyssaceae, 126, 128, 132

O

Oenanthe crocata, 110
Oleaceae, 110
Onopordon acanthium, 153
Orchidaceae, 112, 173, 175
Orchidales, 168, 169

P

Palmae, 172, 173, 174
Palmales, 168
Pandanaceae, 172
Pandanales, 168, 169
Pangium edule, 48
Papaver somniferum, 113

Papaveraceae, 110, 134
Papaverales, 134
Parthenium, 157, 158
Penicillium patulum, 184
  urticae, 184
Petasites, 112, 150, 158, 159, 160, 161
  albus, 150, 158, 159, 160, 161
  hybridus, 150, 158, 159, 160, 162
  japonius, 151, 161
  kablikianus, 151, 159, 160
  paradoxus, 151, 159, 160
  spurius, 151
Petermannia, 177
Phallus impudicus, 180
Phaseolus lunatus, 48
Philydraceae, 174
Piperales, 169
Pittosporaceae, 110, 124–126, 131, 132
Pittosporales, 135
Pittosporum, 124, 125, 156, 131
  buchananii, 126
Plantaginaceae, 131–133
Plantaginales, 131
Plantago, 133
Poales, 169
Polycarpicae, 168
Polygonaceae, 122
Polyporus hispidus, 180, 187, 188
  schweinitzii, 180, 187, 188
  sulphureus, 188
  tumulosus, 183, 185, 185
Pontederiaceae, 174
Poria obliqua, 188
  subacida, 184
Potamogetonaceae, 172, 173
Potamogetonales, 168, 169
Primulales, 131, 133
Prunus, 48, 122
  laurocerasus, 59
Psilophyta, 180
Psilotum nudum, 108
Puccinia graminis, 179, 180, 181
Pyrolaceae, 117

R

Ranales, 134, 168, 176
Ranunculaceae, 123
Ranunculales, 132, 134
Resedaceae, 110
Restionaceae, 174
Restionales, 169
Rhodotorula, 182
  glutinis, 182
  texensis, 182

Rhoeadales, 134
Ricinus communis, 102
Rosaceae, 132
Rosales, 131, 132, 135, 176
Rosiflorae, 126
Rubia tinctorum, 117
Rubiaceae, 117
Rutaceae, 118
Rutales, 131, 132

S

Sambucus nigra, 48
Santalaceae, 110
Santalum acuminatum, 104
Sapindales, 132
Saussurea, 146
　lappa, 152, 155
　pulchella, 155
Saxifragaceae, 124, 132
Saxifragales, 135, 168
Saxifragoideae, 132
Sceletium, 115
Scheuchzeriaceae, 172
Schizophyllum commune, 184, 186
Scirpoideae, 174
Scitamineae, 168
Sedum acre, 116
　sarmentosum, 116
Senecioneae, 140, 144, 145, 149, 150, 154, 158–161
Seriphidium, 156
Smilacoideae, 174
Solanaceae, 116
Solanales, 133
Solanum, 122
Sonchus tuberifer, 147
Sorghum, 48
　vulgare, 49, 63, 173
Spadiciflorae, 168
Spinacia oleracea, 96
Sporobolomyces, 182
　roseus, 181, 182, 183, 184, 186, 187
Stemanaceae, 173
Stereum subpileatum, 180
Steria rhombifolia, 148
Stratiotes, 173
Streptomyces, 189
Strophanthus, 111

T

Tabebuia anellanae, 117
Taxus, 48, 55, 57
Taxodiaceae, 114
Tectona grandis, 117
Tetrahymena, 100
Tmesipteris tannensis, 108
Torricellia, 127
Torricelliaceae, 126
Torulopsis utilis, 100, 104
Tovariaceae, 110
Tricholoma grammopodium, 104
Trifolium repens, 48
Triuridaceae, 173
Triuridales, 168, 169
Tropaeolum, 110
　majus, 64, 65, 66
Tubiflorae, 131, 133, 171
Typhaceae, 172, 173
Typhales, 168

U

Umbelliferae, 110, 116, 122, 125, 126, 132, 135
Umbelliflorae, 126, 131, 134
Urtica, 122
Ustilago hordei, 182, 184, 186

V

Verbenaceae, 117
Vernonieae, 145, 145, 154
Vicia, 48
　angustifolia, 48, 71
　sativa, 67, 68

X

Xanthium, 142, 147
Xyridaceae, 174
Xyridales, 168

Z

Zaluzania augusta, 148
Zieria laevigata, 48, 71
Zingiberaceae, 175
Zingiberales, 168, 169
Zostera, 173

Subject Index

A

Acacipetalin, biosynthetic origin, 71
Acetaldehyde, alanine from, 69
Acetate, precursor of
  acetylenic compounds, 102
  coniine, 116
  fatty acids, 93, 95, 101
  xanthones, 118
  ximenynic acid, 104
Acetone cyanhydrin, dissociation of, 61
  linamarin precursor, as, 63
O-Acetylserine, β-cyanoalanine synthase
  substrate, 68, 69
O-Acetyl serine sulphydrase, 69
Acetyl transacylase, 92
Acetylenic compounds in fungi, 102, 104
ACP, *see* Acyl carrier protein
Active site, β-ketoacyl-ACP synthetase,
  92
  phytoene synthase, 89
Acyl carrier protein,
  plant fatty acid synthesis and, 92
  prosthetic group of, 91–92
Acyl lipid—fatty acid synthesis, coupl-
  ing, 101–102
Adonitol, accumulation of, 122
Aglycones, of cyanogenic glucosides, 47,
  48
Alanine, enzymic synthesis of, cyanide
  and, 69
Aldoximes, glucoside biosynthesis, in,
  58, 62, 64–65
Algae (*see also under* Blue-green *and*
  Green)
  unsaturated fatty acid formation in,
  95, 96–97, 98, 99
Alizarin, from shikimate, 117
Alkaloids (*see also specific names*) 111,
  112, 115, 116, 122, 123, 130, 139, 158
Almonds bitter, hydroxynitrile lyase, in,
  49–50
  sweet, β-glucosidases, in, 49
Aluminium, accumulation of, 122, 123

Aluminium chloride, in flavonoid analysis, 2, 17, 18
Ambrosanilides (*see* Pseudoguaianolides)
Amino acids (*see also specific names*)
  biosynthesis *via* nitriles, 69
  fossils, from, 108
  oxidation of, 66
  precursors of
    aglycones of thioglucosides, 63–66
    asparagine, 72
    cyanogenic glycosides, 51, 53, 54,
      55–57, 58, 59, 60, 62, 70, 71
    mustard oil glycosides, 63–66
Amino cyanobutyric acid, as glutamate
  precursor, 69
2-Amino propionitrile, in cyanide metabolism, 69
Anabaena, unsaturated fatty acid forma-
  tion in, 195
Anabasine, biosynthesis of, 116
Anaerobic conditions, fatty acid synthesis
  under, 94, 95–96
Aniba, 6-phenylcoumalin and paracotoin
  from, 108
Annona reticulata, reticuline from, 113
Anthemideae, sesquiterpene lactones in,
  145, 146, 154, 156
Apigenin, UV spectra of, 12, 19
Arctotideae, sesquiterpene lactones in,
  145, 146
Ariales, chemical relationships of, 126,
  127, 132
Ascomycetes, polyketides, and, 184
Asparagine, metabolism, 67–68, 71–72
Aspidosperma, alkaloids in, 111
Asteraceae, sesquiterpene lactones in,
  145, 146
Athrotaxin, origin of, 114, 115
ATP (Adenosine triphosphate),
  biosynthesis, in, 68, 84, 93
  decline of, in pea, 201
ATPase, 201, 203

Aucubin, in,
  Cornales, 127, 128, 129, 130
  Hippuridaceae and Plantaginaceae, 133
Avocado, fatty acid synthesis in, 91, 95

B

Bacteria, tyrosine degradation by, 184
Barley seedlings, cyanide metabolism by, 66
Basidiomycetes, 179–191
  ammonia lyases in, 181
  cell wall components in, 188
  cinnamic acid metabolism in, 179–191
  phenylalanine metabolism in, 182–183, 183–187
Bathochromic shifts, in flavonoid analysis, 13, 14, 15, 17, 20, 22, 27, 31
Batis, classification of, 124
Bayin, UV spectra of, 19
Bean leaves, cycloartenol synthesis in, 89
Benzaldehyde, dhurrin and, 49, 50
Benzyl glucosinolate, precursor of, 63, 64, 65
Benzylisoquinoline, from tyrosine, 113
Betalains, occurrence of, 111, 112, 124
Bile acids, "primitive", 108
  salts, 112
Biochemical systematics, use of term, 108
Biosynthesis and molecular taxonomy, 112–118
Bisabolane skeleton, biosynthesis of, 144
Blue-green algae, fatty acids in, 95
Blue lupin,
  asparagine precursors in, 72
  synthases in, 68, 69
Bryophyta, cinnamyl compounds in, 180

C

Caffeic acid, in Cornales, 127, 128, 129
Calenduleae, sesquiterpene lactones in, 145, 146
Cardenolide glucosides, occurrence of, 110
Cardinane skeleton, biosynthesis of, 144
Carotenes, α-, β-, γ-, δ-, ϵ- and ζ-, 78–82, 84
  biosynthetic pathway for, 83
Carotenes, cyclic, formation of, 78–82
Carpobritis chilense, fatty acid metabolism in, 98

Carrot root, carotene synthesis by, 80, 82
  squalene and phytoene synthesis by, 79, 80, 82
Centrospermae, phytochemistry and classification of, 124
Chamazulene, chamazulenogenic substances, 140, 162
"Chemical races", 162
Chemotaxonomy, origin and use of term, 107, 108
  sesquiterpenoids in Compositae, of, 139–165
Cherry laurel, biosynthesis in, 53
  prunasin precursors in, 58, 59
Chlorella, cyanide assimilation by, 67
  unsaturated fatty acid formation by, 95, 98, 99, 100, 101
Chlorogenic acid (*see also* Caffeic acid)
  Cornales, in, 127
  Pittosporaceae, in, 125
Chloroplasts, fatty acid synthesis in, 93, 95
  phytoene synthesis by, 84–85
  stearoyl-S-ACP desaturation by, 96
Cichorieae, sesquiterpene lactones in, 145, 147
Cicuta virosa, polyacetylenes in, 110
Cinnamate, cinnamic acid, metabolism of, 179–191
Cinnamyl compounds,
  characteristic of higher plants, 180
  formation of complex derivatives of, 187–188
  synthesis of, 179–180
  transformation of, 182–183
Cinnamyl ring, of flavonoids, 13
Classification of plants, phytochemistry of, 123–134
Claviceps purpurea, ricinoleic acid formation in, 102
Clostridia, unsaturated fatty acid formation in, 94
Clover, red, cyanide incorporation by, 66
  white, cyanogenic glucoside biosynthesis in, 53
Coelacanth, "primitive" bile acids of, 108
Coenzyme A, 91, 92, 93
Colchicine biosynthesis, 114–115
Colchicum, colchicine biosynthesis by, 114
Compositae, bitter and toxic principles in, 140, 146
  chemotaxonomy of sesquiterpenoids of, 139–165
  classification of, 139

Compositae, *cont.*—
  Magnoliaceae and, 141
  tribes of, 144, 145
Conifers, patterns of constituents of, 111
  pinoresinol in, 109
Coniine biosynthesis, 116
Cornaceae, phytochemistry and classification, 126–131
Cornales, Araliales, relation with, 127
  cornin in, 127, 128, 129
  families in, 126
  iridoid glucosides in, 127, 128
Corynebacterium diphtheriae, fatty acid
  dehydrogenation in, 98
p-Coumaric acid, as enzyme substrate,
  181
Crepis rubra, crepenynic acid formation
  in, 104
Cyanide (HCN),
  asparagine, incorporation into, 66–67,
    68, 71, 72
  cyanogen glycoside precursor, as, 49,
    50, 52, 60, 61, 62
  enzymic reactions involving, 49, 68, 69
  mandelonitrile metabolism, inhibition
    of, 59
  metabolism of, 50, 52, 66–70, 71, 72
  microbial metabolism of, 68, 69, 70
  production, 47, 51, 70
β-Cyanoalanine, cyanide metabolism
    and, 67, 68, 70, 71
  enzymic formation of, 68
  phenylalanine as precursor of, 71
β-Cyanoalanine synthase, 68, 69
Cyanogenesis, by fungi, 70
  by plants, 47
Cyanogenic compounds, 122
Cyanogenic glycosides,
  aldoximes and, 58, 62, 65
  biosynthesis of, 47–74
  catabolism of, 49–50
  occurrence of, 47, 48, 70
Cyanohydrins, cyanogenic glycosides
  and, 47, 50, 58, 59, 60, 61, 62
Cyanophoric plants, 71
Cycloartenol, mechanism of formation,
  85–89
  occurrence of, 87
  squalene, formation from, 86, 88, 89,
    116
Cyanareae, classification of, 154–156
  germacranolide distribution in, 152
  sesquiterpene lactones in, 145, 146,
    154–156

Cysteine, cyanide and, 68, 69
L-Cysteine hydrogen sulphide lyase, 68
Cysteine synthase, in higher plants, 69
Cytochromes, molecular taxonomy and,
  118

D

Decanoate, fatty acids from, 93, 97
Deoxylapachol, origins of, 117
Dhurrin, aglycone, origin of, 51, 55, 70
  biosynthesis of, in sorghum, 49–52, 63
  catabolism of, 49–50
  precursors of, 51, 52, 55–57, 60–61
Dicotyledons, flavonoids in, 169–172
  Monocotyledons and, 172
Dihydroflavonols, 2
  UV spectra of, 20–31
Dihydroquercetin, UV spectra of, 21, 26

E

Ellagic acid, in Cornales, 126–129, 131
Emodin, polyketide origin, 117
Endocrocin, polyketide origin, 117
Enoyl-ACP reductase, 92
Enzymes (*see under individual names*)
Equisetophyta, cinnamyl compounds in,
  180
Eremophilanes, furans from, 140
Eriodictyol, UV spectra of, 21, 26, 30
Escherichia coli, acyl carrier protein of,
  91, 92
  cyanide assimilation by, 67, 68
  β-cyanoalanine formation in, 68
Essential oil constituents, 122, 139, 155
Essential oils, in Pittosporaceae, 125, 131
Euglena, desaturase system of, 101
  fatty acid formation in, 95, 96
Eupatorieae, sesquiterpene lactones of,
  145, 148, 160
Euphorbia latex, lanosterol formation in,
  87, 89

F

Farnesyl pyrophosphate, geranyl pyrophosphate, from, 76–77
  squalene precursor, as, 77
Fatty acid(s) (*see also under individual
  names*)
Fatty acid biosynthesis,
  acetylene bond, introduction of, 104
  plants in, 91–105
  saturated, 91–93
  temperature effect on, 104

Fatty acid biosynthesis *cont*—
  unsaturated,
    chain length and desaturation, 98
    desaturase enzyme systems for, 96,
      98, 100, 101, 104
    mechanisms of formation, 93–94
    monoenoic, 93–98
    pathway in leaves, 95
    polyunsaturated, 100–101
    unusual, 102–104
Ferredoxin, fatty acid desaturation and,
  96
Flavanones, 2
  skeleton of, 22
  UV spectra of, 20–31
Flavone glycosides, 2
  UV spectra of, 2–20
Flavonoid aglycones and glycosides,
  NMR analysis of, 31
Flavonoids (*see also under names of sub-
  groups and individual names*)
  analysis of structure by spectral
    methods, 1–45
  Basidiomycetes, absence from, 189
  biosynthesis of, process related to, 188
  conifers, in, 111
  content,
    pea tendrils, in, 202
    growth, in relation to, 193, 197, 198
  Dicotyledons, in, 169–172
  enzyme activity, effect on, 196–197
  geographical distribution of *Crocus*,
    and, 175
  Monocotyledons, in, 172–175
  photomorphogenisis and, 193–204
  polyketide biosynthesis and, 181
  precursors, light effect on, 198–200
  taxonomic implications of, 174, 175–
    177
Flavonols, structure and nomenclature
  of, 2, 13
  Pittosporaceae, in, 125
  UV spectra of, 2–20
Flax,
  plants; linamarin biosynthesis in, 57–
    58, 62–63
  seedlings, cyanide metabolism of, 50,
    61, 66
  seeds, fatty acids of, 104
  shoots, linamarin precursors in, 54–55,
    59, 61
Fossils, organic compounds from, 108
Fungi, acetylenic compounds in, 102
  cinnamyl compounds in, 179–180

  cyanide metabolism in, 69–70
  Imperfecti, ammonia lyases in, 181
  lignins and, 188
  secondary metabolite synthesis in, 179

G

Galangin, NMR spectroscopy of, 32
Gallic acid, in Cornales, 127, 128, 129,
  131
Genistein, UV spectra of, 21, 28
Geranyl pyrophosphate, formation of,
  76–77
Geranylgeranyl pyrophosphate, meta-
  bolism, of, 77, 83
Germacranolides,
  Compositae in, 142, 143, 145, 146
  Cynareae, in, 152, 156
  Heliantheae, in, 148
Ginkgo biloba, ginkgolides from, 108
Glugonasturtiin, biosynthesis, 64
β-Glucosidase, salicin as substrate of, 49
β-Glucosidases, cyanogen hydrolysis and
  49
Glucosinolate compounds (*see also*
  Mustard oil glucosides)
  biosynthesis of, 63–66
  families producing, 110
Glucotropaeolin, precursor of, 63
Glutamic acid, cyanide incorporation
  and, 69
Glutamine biosynthesis, 71–72
γ-Glutamyl transferase, cyanide and,
  67–68
Glycosides, *see also under* Cyanogenic
  cardenolide, 110–111
Glycosyl transferase, limanarin biosyn-
  thesis and, 63
Green algae, fatty acid and desaturation
  by, 95, 96, 98, 99
Guaianolides, 142, 143
  Ambrosiaceae, in, 148
  Compositae, in, 145, 146, 147
  Cyanareae, 154

H

Haemoglobins, molecular taxonomy
  and, 118
HCN (*see under* Cyanide)
Hedacidin, biosynthesis of, 66
Hedera helix, leaf constituents of, 126
Helenieae, distribution of, 147, 154
  sesquiterpene lactones in, 145, 165–158
Heliantheae, lactones in, classification
  and, 144, 145, 147, 149, 154, 156, 158

Hesperidin, trimethyl silylation of, 33
Hexadecenoic acid, biosynthesis of, 99–100, 102
Hexadecenoyl-phosphatidyl glycerol, biosynthesis of, 100
Hexitols, accumulation of, 122
Hippuridaceae, classification of, 133–134
Hippuris, chemical characters of, 133, 134
Homogyne, sesquiterpene lactones of, 161
β-Hydroxyacyl-ACP dehydrase, 92
N-Hydroxyamino acids, biological role of, 66
p-Hydroxybenzaldehyde, metabolism of, 52
α-Hydroxyisobutyric acid, linamarin and, 53
α-Hydroxyisobutyronitrile, linamarin precursor, 59, 60
p-Hydroxy-L-mandelonitrile, enzymic formation, 49
Hydroxynitrile lyase, in sorghum and almonds, 49–50
α-Hydroxynitriles, cyanogenic glucosides and, 47, 58, 62
α-Hydroxyphenylacetonitrile, as prunasin precursor, 59, 60
p-Hydroxyphenyl-DL-lactic acid, as dhurrin precursor, 52
Hydroxysugiresinol, athrotoxin and, 114, 115
Hypsochromic shifts, exhibited by flavonoids, 14, 20, 27

I

Indoleacetic acid oxidase, 194, 195, 196, 197
Ionone rings, of cyclic carotenoids, 78–82
Inuleae, sesquiterpene lactones of, 145, 148, 149
Iridoid glucosides (*see also under specific names*), 126, 129
    Cornales, in, 128, 129, 131
    Plantaginaceae, in, *133*
Isocitric acid, accumulation of, 122
Isobutyraldoxime, as linamarin precursor, 57, 58, 59, 63, 64
    isopropylglucosinolate precursor as, 14, 15
Isobutyronitrile, as linamarin precursor, 59, 60
Isoflavone skeleton, numbering system, 22

Isoflavones, UV spectra of, 22–31
Isolated chloroplasts (*see under* Chloroplasts)
Isoleucine, as biosynthetic precursor, 53, 54, 70, 71
Isopentenyl pyrophosphate metabolism, 75, 76
Isoprene precursors, 75
Isopropyl glucosinolate, precursors of, 63, 64, 65

K

α-Keto acid oximes,
    amino acid oxidation to, 66
    nitrites, conversion to, 54, 65
β-Ketoacyl-ACP, reductase and synthetase, 92
α-Ketoisovaleric acid oxime, as linamarin precursor, 57–58
α-Ketoximes (*see* α-Keto acid oximes)

L

Lanosterol, as sterol precursor, 85, 86, 87, 88, 116, 117
    in *Euphorbia* latex, 87, 89
Latimeria bile acids, "primitive", 108
Laurate, leaf fatty acids from, 97
Lignins, formation of, 181, 187–188
    absence from fungi, 188
Linamarin, biosynthesis, 47, 53, 54, 55, 57–58, 59, 60, 61, 62–63, 64, 70
    occurrence, 48, 49, 71
Linen flax (*see also under* Flax *and Linum usitatissimum*)
    linamarin biosynthesis in, 49
Linoleic acid, biosynthesis, 96, 100, 103
    ricinoleic acid precursor, as, 102
Linolenic acid, biosynthesis, 103
Litorella uniflora, similarity to *Plantago*, 133
Lotaustralin, biosynthesis, 47, 53, 54, 70, 71
    occurrence, 48, 70, 71
Lotus seedlings, linamarin and lotaustralin in, 71
Luteolin, spectral data for, 12, 16, 19, 36
Lycopophyta, cinnamyl compounds in, 180
Lycopene, biosynthesis, 78, 79, 80, 82, 83, 84
Lysine, in alkaloid biosynthesis, 116

M

Magnoliaceae, and Compositae, similarities, 141

S-Malonyl CoA, in fatty acid synthesis, 91, 92, 93

Malonyl transacylase, 92

Mannitol, accumulation of, 122

Mastixia, iridoid glycosides in, 127, 130

Mentha, terpene production by, 113

O-Methylthreonine, linamarin synthesis inhibitor, 63

Mevalonic acid (MVA),
anthraquinone precursor, as, 117
carotenoid and triterpenoid biosynthesis, in, 75–85, 87

Microbial metaboilsm, of cyanide, 68, 69, 70

Molecular biology, and molecular taxonomy, 118

Molecular taxonomy,
biosynthesis and, 112–118
development in, 107–120
molecular biology and, 118–119
use of term, 108

Monocotyledons,
classification of, 167, 168, 169
dicotyledons and, 172
flavonoids of, 172–175
leucoanthocyanins in, 177
origin of, 176
"reticulate" taxanomic structure in, 176

Monoenes, saturated fatty acids and, 98

Monoenoic acids,
biosynthesis, 93–98
metabolism to di- and trienoic acids, 101

Morphine, biosynthesis, 113, 114

Mustard oil glucosides (*see also* Glucosinolates)
biosynthesis, 63–66

Mustard oils, screening for, 110

Mutisieae, sesquiterpene lactones in, 145, 146

Myristate, myristic acid,
acetate as precursor, 101
fatty acids from, 97, 98

N

Natural products, compilations of, 109

Neurosporene,
biosynthetic pathway for, 83
phytoene desaturation to, 78

Nicotiana, nicotine biosynthesis in, 116

Nicotine, biosynthesis, 116
species containing traces of, 122

Nitrile moiety of cyanogenic glucosides, origin, 51, 54, 58–59, 62

Nitriles,
amino acid precursors in fungus, as, 69
cyanogenic glucoside precursors, as, 60, 65
keto acid oximes, conversion to, 53, 54, 65

NMR (Nuclear magnetic resonance),
spectroscopy of flavonoids, 1, 31–43
solvents for, 31

O

9-Octadecen-12-ynoic acid, from oleic acid, 103, 104

11-Octadecen-9-ynoic acid, from acetate, 104

Octanoate, octanoic acid, fatty acids from, 93, 97

Oenanthe, polyacetylenes in, 110

Oleic acid,
acetate from, 96, 101
anaerobic formation, 94
crepenynic acid precursor, 104
linoleic acid precursor, 100
ricinoleic acid precursor, 102, 103

Oleyl-S-CoA,
formation, 93
ricinoleic acid formation, substrate, 102

Organic geochemistry, 108

Orobol, UV spectra of, 26, 30
-7-O-glucoside, UV spectra of, 30

Oximes, cyanogenic glycoside precursors, as, 57, 62–63

Oxygen-dependent fatty acid desaturation, 93, 94, 95, 100, 102, 104–105

P

Palmitate, palmitic acid,
acetate, from, 96, 101
hexadecanoic acid precursor, 99, 100
lipids in leaves from, 97

Papaveraceae, phytochemistry and classification, 134

Paracotoin, from *Aniba*, 108

Parkeol, enzymic formation of, 88

Pea, effect of light on growth, 194–195
epicotyls, response to red light, 194–200

Pea, *cont*—
plumules, flavonoids and, 197
seedlings, cyanide and asparagine in, 66
light effects on, 193
tendrils, flavonoids in, 200–203
response to touch of, 200–201
Peach seedlings, prunasin biosynthesis in, 53
Pentitols, accumulation of, 122
Petasites, sesquiterpenes from, 112, 159
Phenolic acids, in Cornales, 128
Phenylacetaldoxime,
benzyl glucosinolate precursor, as, 64, 65
N-hydroxyphenylalanine as precursor of, 66
prunasin precursor, as, 58
Phenylacetonitrile, as prunasin precursor, 59, 60
Phenylalanine, as precursor of,
benzylglucosinolate, 63–64, 65
β-cyanoalanine, 71
γ-phenylbutyrine, 64
prunasin, 53, 54, 58, 59, 60
Phenylalanine ammonia lyase, 182, 189
γ-Phenylbutyrine, as gluconasturtiin precursor, 64
6-Phenylcoumalin, from *Aniba*, 108
Phenylpropane carbon skeleton, 180
Phenylpyruvic acid, as prunasin precursor, 58
Photomorphogenesis and flavonoids, 193–204
Phylogenetic relationships, protein structures and, 118
Phytochemical data, limitations of, 161–162
Phytochemistry, comparative, 109–112, 119, 122, 135, 163,
plant classification and, 123–134
Phytochrome, 193, 195, 197–198, 201, 203
Phytoene, biosynthesis, 75–85
formation and desaturation, 82–84
Phytoene synthase, 82
Phytofluene, as carotenoid intermediate, 83, 84
Phytosterols, 85–89, 116
Pigment synthesis, 193
Pinoresinol, in conifers, 109
Pittosporaceae, phytochemistry and classification of, 124–126, 131, 132

Pittosporum sp., comparison of leaf constituents, 126
Plant classification, phytochemistry and, 123–124
Plant movements, 200–201
Plantaginaceae, phytochemistry and classification of, 131–133
Plantago, chemical studies of, 133
Polyacetylenes, Compositae and Umbelliferae, from, 109–110
Pittosporum, from, 122, 125
Polyenoic acids, biosynthesis, 103, 104
Polyunsaturated fatty acids, biosynthesis, 100–101
stereochemistry of, 101
Potato leaves, triterpenes from MVA in, 87
Prunasin, biosynthesis, 53, 54, 58, 59, 60, 70
zierin precursor, as, 71
Pseudoguaianolides (*see also* Ambrosanilides), 142, 143, 145, 147, 148, 156, 158, 162
Compositae, in, 145
Heliantheae, in, 148, 156
Psilophyta, cinnamyl compounds in, 180
Psilotum, psilotin from, 108

Q

Quinic acid,
accumulation of, 122
Cornales, in, 128
Pittosporaceae, in, 126
Quercetin, UV spectra of, 19
Quercitrin, UV spectra of, 12, 16

R

Rape seed, temperature and fatty acids of, 104
Red clover, cyanide metabolism by, 66
Ricinoleic acid, biosynthesis, 102, 103
Ricinus, ricinoleic acid formation in seeds of, 102
Rubber, from MVA, 78

S

Sakuranin, UV spectra of, 30
Salicin, as glucosidase substrate, 49
Santalum, ximenynic and formation by, 104
Santanolides (*see also* Eudesmanolides), 142, 143, 146, 147, 156,
Ambrosiaceae, in, 148
Compositae, in, 145

Saponins, in *Pittosporum*, 125, 131
Sceletium, alkaloids from, 115, 116
Senecio alkaloids, 158
Senecioneae, 140, 144, 149, 154
  eromophilane derivatives, in, 150
  sesquiterpene lactones and classi-
    fication, 145, 158–161
Serine, cyanide reaction, 68
Serine sulphydrase, cyanide metabolism
  and, 67
Sesquiterpene lactones,
  biosynthetic pathway, 142–144
  classification of Compositae, and, 144–
    158
  definition as taxonomic characters,
    140–142
  types of, 142
Sesquiterpenoids (*see also individual
  names*), 139–165
Shikimic acid,
  accumulation of, 122
  precursor of,
    alizarin, 177
    dhurrin, 51
    xanthone, 118
Silicic acid, accumulation of, 122
Sodium acetate, as reagent in flavonoid
  analysis, 2, 15–17, 43, 44
Sodium methoxide, as reagent in flavonoid analysis, 2, 15, 28, 43, 44
Sorbitol, accumulation of, 122
Sorghum (*see also Sorghum vulgare*)
  cyanide metabolism in, 50, 61, 66
  dhurrin biosynthesis in, 49, 51, 55, 63
  dhurrin catabolism in, 49, 63
  hydroxynitrile lyase in, 49, 50
Spinach, fatty acid metabolism in, 93, 96
Spinacia oleracea (*see under* Spinach)
Sporophores,
  cinnamic acid in, 179
  hispidin in ripening of, 188
Squalene,
  biosynthesis, stereospecific, 75–78
  cyclisation to lanosterol and cyclo-
    artenol, 86, 88
  farnesyl pyrophosphate as precursor,
    77
Squalene 2,3-oxide as sterol precursor,
  85, 89, 116
Stachyose, accumulation of, 122
Stearic acid, from acetate, 96, 101, 103
  unsaturated fatty acids from, 103
Stearoyl-*S*-CoA, dehydrogenation of,
  93, 96

Sterculic acid, 99, 100
Sterols, from squalene, 85, 86, 88
Strophanthus, cardenolide glycosides of,
  111
Structure analysis, by NMR spectroscopy, 1, 31–43
  by UV spectroscopy, 1, 2–31
Succinic semialdehyde, glutamate from,
  69
Sunflower seeds, temperature and fatty
  acids in, 104

T

Taxa, screening for compounds of interest in, 109–112
  chemical evidence for classification of,
    121–138
  classification of, comparative studies
    and, 134
Taxiphyllin, biosynthesis, 55, 57, 70
Taxus, taxiphyllin synthesis in, 55, 57
Tectoquinone, origin of, 117
  analogues of, 118
Temperature, effect on fatty acid biosynthesis of, 104–105
Terpenoids (*see also under specific names*),
  111, 139
Tetradecenoate, fatty acids in leaves
  from, 93
Tetrahymena, polyunsaturated acid in,
  100
Texasin, UV spectra of, 26, 30
Tmesipteris tannensis, psilotin from, 108
del-Tomatoes, carotenoid polyene syn-
  thesis in, 79, 80, 81, 84
Torulopsis utilis,
  polyunsaturated acid formation in, 100
  stearoyl-*S*-CoA desaturation rate in,
    104
Trimethylsilyl ether derivatives of flavonoids, for NMR spectroscopy, 31, 32–43
Triterpenes, 111, 122, 125, 131
  biosynthesis of, 75–90
Tropaeolum, glucosinolates in, 110
Tropolones, in conifers, 111, 114
Tyrosine, as precursor of,
  dhurrin, 51, 54, 55–57
  morphine, 113
  taxiphyllin, 55, 57
Tyrosine ammonia lyase, 181, 182

U

UDP glucose, in linamarin formation,
  63

Umbelliferae, polyacetylenes in, 110
Unsaturated fatty acids, desaturation reaction forming, 93, 95
Uredospores, cinnamic acid in, 179
tyrosine ammonia lyase of, 181
UV spectra (*see also under names of specific compounds*)
flavones and flavonols of, 2–20
isoflavones, flavanones and dihydroflavonols, 20–31

V

Valine,
isopropyl glucosinolate precursor, as, 64, 65
linamarin and lotaustralin precursor, as, 53, 54–55, 60, 63, 70, 71
Verbascose, 122
Vernonieae, sesquiterpene lactones of, 145, 146, 154
Vetch, common cyanide metabolism by, 67
Vicia sativa, cyanide metabolism by, 67, 68

Vicianin, β-cyanoalanine precursor, as, 71

W

Watercress (*see also Nasturtium officinale*)
gluconasturtiin in, 64
Wheat rust uredospores, 179
White clover, cyanogenic glucoside biosynthesis in, 53
cyanide incorporation in, 66

X

Ximenynic acid, from acetate, 104

Y

Yeasts, fatty acid desaturation in, 93, 96, 100

Z

Zeacarotenes, in carotenoid biosynthesis, 80, 83
Zierin, origin of, 71
Zoochemistry, comparative, 109–112